The Patrick Moore Practical Astronomy Series

Series Editor
Gerald R. Hubbell
Mark Slade Remote Observatory
Locust Grove, VA, USA

The Patrick Moore Practical Astronomy Series is a treasure trove of how-to guides for the amateur astronomer. The books in this series are written for hobbyists at all levels, from the enthusiastic newcomer to the veteran observer. They thus go far beyond more general, popular-level books in both scope and depth, exploring in detail the latest trends, techniques, and equipment being used by amateur astronomers around the world.

You will find herein a diverse list of books on constellations, astronomy catalogues, astrophotography, eclipse chasing, telescope equipment, software, and so much more. All books in the series boast full-color images as well as practical sections for putting your newfound knowledge to use, including star charts and target objects, glossaries, hands-on DIY projects, troubleshooting walk-throughs, and a plethora of other helpful features.

Overall, this series bridges the gap between the many introductory books available and more specialized technical publications, providing digestible, hands-on guides for those wishing to expand their knowledge of the night skies.

Richard Stember

Share the Universe

A Guide to Outreach Astronomy

 Springer

Richard Stember
Science Heads Inc.
Lake Forest, CA, USA

ISSN 1431-9756   ISSN 2197-6562   (electronic)
The Patrick Moore Practical Astronomy Series
ISBN 978-3-031-53494-2   ISBN 978-3-031-53495-9   (eBook)
https://doi.org/10.1007/978-3-031-53495-9

Cover credit: Mr. Zach Schierl

This Springer imprint is published by the registered company Springer Nature Switzerland AG
The registered company address is: Gewerbestrasse 11, 6330 Cham, Switzerland

Paper in this product is recyclable.

This book is dedicated to my wife, who puts up with all my passions, ideas, and projects. She gives me the space and encouragement to pursue my passions. Without her, my life would contain far less love and meaning. And I dedicate this book to all the amateur astronomers who labor on weeknights and weekends sharing their passion for astronomy with the public. You inspire me with your knowledge, dedication, and sense of purpose. Your efforts are immensely appreciated and should never go unnoticed.

Preface

Inspiration can hit you at any time. Sometimes it feels like a brick, other times just a gentle breeze. It was a cool cloudless autumn evening when inspiration swept gently across my face. Octobers can be hot in Southern California, but on this night it was cool and clear; perfect for a star party. It didn't take much effort to convince my 11-year-old son to walk down the street with me to his elementary school. His school was hosting its annual Family Astronomy event. It had been a long time since I had looked through a telescope, and I was probably even more excited than my son.

We arrived at the back of his school an hour after sunset. Arrayed before us were a dozen or more telescopes of various types. A sizable crowd of parents and students had organized themselves into lines. The telescopes ranged in size from a large 25-inch truss Dobsonian, which required climbing a ladder to look through the eyepiece, to compact but seemingly more advanced computer-controlled telescopes mounted on tripods. There was even a rotating, articulated chair with binoculars attached ready to scan the heavens.

The air was thick with excitement, and we could hear animated conversations punctuated with "Wow!" or "That's amazing!" Children were jumping up and down and running from line to line. The sky was dark, and time was precious. Bedtime was fast approaching, and yet there was so much to see.

I knew that this was a special event – something that my son and I would remember for a long time. That night, at the age of 42, I was introduced to the world of Outreach Astronomy. The event inspired me to purchase my first telescope, join a local astronomy club, and start my journey that continues to this day sharing astronomy with the public. Several years on, I would form a non-profit dedicated to supporting STEM education and raising science literacy. That organization, Science Heads Inc., has since hosted hundreds of events reaching thousands of children and adults. It has also expanded to multiple states, adding volunteers, and forming chapters stretching clear across the country.

I am now at the age when many people retire. But my outlook on life has only increased my passion. I look forward to many more years of sharing what I know with people who, like me, are curious about the Universe. It's a noble passion – one which provides direction and purpose.

Over the years, I have been very fortunate to meet and work with many like-minded amateur astronomers. They taught me much more than just how to use a telescope. With their guidance, I learned how to explain complex concepts to those new to astronomy. I learned how to put myself in the shoes of a person unaccustomed to looking through a telescope; how to work with children, adults, physically challenged individuals, and those who are developmentally delayed.

Astronomy is the most approachable of all the sciences. It's exciting and easy to understand. It's also easy to participate in making it a perfect vehicle to explain how science "works." All it takes is to look up. Lack of knowledge, intellectual or physical abilities, or fully functioning senses need not be obstacles to learning about the Universe.

This book is a guide for hosting many types of astronomy outreach events. In it, I will share some of the best practices, tools, and resources available. It will highlight the important work being done by leaders in the field and will explain how you can reach an ever-broader audience while serving your community.

If you have a passion to share knowledge like I do, this guide will help you better plan, run, and participate in outreach events. You will learn how to engage people and include many who have historically been left out or ignored.

The time that I have spent doing outreaches has more than been rewarded with the satisfaction that I am helping raise science literacy in my community. To this day I still enjoy hearing "Wow!" when someone sees Saturn or the Moon for the first time in a telescope. And I enjoy having a guest walk away feeling like they just learned something new.

As you would expect, my own learning continues. I welcome your suggestions on the topics covered in this book. Feel free to email your comments to me at richard@scienceheads.org. If appropriate and possible, I may include them in subsequent editions of this guide.

Lake Forest, CA, USA   Richard Stember

Acknowledgments

Over the 25 years that I've been doing outreach, I have been honored to work with dozens of people who graciously and selflessly shared their expertise. Their passion and generosity are truly inspiring and infectious.

Since it is not possible to thank them all on these pages, I will instead recognize the people and organizations who provided information and materials helpful to writing this guide. Without their help this book would not have been possible.

I am indebted to Mr. James Benet who many years ago introduced me to outreach astronomy and provided helpful advice while writing this book; Ms. Noreen Grice who, through her books and correspondence, educated me about working with differently abled audiences; Mr. Zachary Schierl for sharing his experience at the National Park Service and insight about informal interpretation; Mr. John Land of the Tulsa Astronomy Club who opened many doors for me to collaborate with members of his club; Ms. Peggy Walker of the Broken Arrow Sidewalk Astronomers and a local Chapter Coordinator of Science Heads Inc. for sharing ideas about how to work with intellectually challenged audiences; Ms. Rebecca Hammond and Mr. James Hammond for sharing the hands-on activities that they developed for Science Heads Inc.; and Ms. Genaye Channel of Girl Scouts USA for providing important information about the Girls Scouts.

I also wish to thank the staff at Springer Nature and Mr. Michael Maimone for their help in making this book a reality.

Contents

List of Figures

About the Author

Richard Stember has been involved in Outreach Astronomy for nearly 25 years. He started as a volunteer with the *Orange County Astronomers*, then founded a non-profit organization called *Science Heads Inc.* This was not the first time Richard ventured into running a non-profit. For 16 years, he was a board member and president of *Comprehensive Child Development Inc.*, a Long Beach California-based organization that helps low-income families. Richard received a bachelor's degree in chemistry from the State University of New York, College at Oneonta. After graduation, he went to work for several analytical instrumentation manufacturers where he provided technical support to scientists and researchers working in pharmaceutical, consumer products, and environment testing laboratories. Over the years, he has been employed as a Product Specialist, Marketing Product Manager, Sales Engineer, Programmer, and CEO. In 1976, he set out on his own and founded LABTRACK, a software company that develops scientific data management software. Richard wrote and introduced the world's first commercial Electronic Lab Notebook software. This and other software he developed are used by hundreds of companies around the world to improve their research, R&D, and quality control programs.

Chapter 1
Introduction to Outreach Astronomy

A teacher affects eternity; he can never tell where his influence stops.

Henry Adams

Everyone finds themselves in the role of educator at some point in their life. Parents are educators for their progeny. Employees are often asked to take a new employee under their wing. Good managers lead by example. And professionals of all stripes help their clients by dispensing information pertaining to their field of expertise. Sharing knowledge with others is an essential part of being human. We naturally feel the need to educate and enlighten others.

Education itself can be categorized into two distinct types: *Formal Education* and *Informal Education*. The United Nations Educational, Scientific and Cultural Organization (UNESCO), describes formal education as *"… institutionalized, intentional and planned through public organizations and recognized private bodies…"*.

Many people are unfamiliar with the term *Informal Education*. It occurs outside of or after formal education has ended. Visiting a museum is engaging in informal education. Watching documentaries, reading books, and attending lectures are also forms of Informal Education. Informal Education differs from Formal Education not just in when and where it occurs but also by content and learning process. Even the objectives, motivations, and interests of participants can be significantly different.

1.1 Outreach Astronomy as Informal Education

Describing oneself as a "lifelong learner" is common today. Individuals often actively seek out sources of information and opportunities to learn about subjects that are of interest. Sources may include books, magazines, web sites, lectures, museums, nature centers, and documentaries.

People also passively assimilate information that is presented to them. It may be from a TV show, commercial, movie, or a social media site they just visited; they

© The Author(s), under exclusive license to Springer Nature Switzerland AG 2024
R. Stember, *Share the Universe*, The Patrick Moore Practical Astronomy Series,
https://doi.org/10.1007/978-3-031-53495-9_1

may not have been intentionally seeking information, but it is assimilated none-the-less. With so many sources available credibility becomes an issue.

Frequently people find it hard to determine the quality and veracity of information. Should the source be trusted or not? Is the information factual or fiction?

If you are seeking information, then presumably you would choose a quality source. Looking up the definition of a word in a published dictionary would seem more reliable than asking a stranger. If the quality of the source is known, then one can have more confidence in the information provided.

The *Informal Educational Landscape* that is part of our culture is rich with wonderful museums, science centers, and libraries. These are trusted sources of information. But the landscape is also rife with an almost unlimited number of questionable sources. This includes entertainment thinly disguised as educational or newsworthy. These sources can easily be misunderstood to be factual when they are not.

We also live in a commercialized and politicized world. There are few aspects of modern life that are not influenced by individuals, companies, and organizations vying for our attention, our money, our vote, and our loyalties. Television commercials, for example, are produced to inform the public about the virtues of a specific manufacturer's product. Considering the motivation involved, how can we be sure of the veracity of the information? Is it truthful, partially true, misleading, or entirely false?

This matters because recent studies show most of what people "know" about science comes from informal educational sources. In a research paper titled "The 95 Percent Solution", Oregon State University professors John Falk and Lynn Dierking calculated that a typical American spends less than 5% of their lifetime in school. They estimated that most of what people know about science is learned elsewhere. According to the authors, formal education is one of the least significant sources of science knowledge.

Is it no wonder that people are frequently misinformed about science, come to accept unscientific ideas, and subscribe to hoaxes and conspiracy theories? The informal education landscape makes it easy. And there is no shortage of people eager and motivated to mislead and misinform for their own benefit.

1.2   A Misinformed Public

In 1998 Touchstone Pictures released the popular action movie *Armageddon*. The movie grossed over $ 200 million in the United States and $ 550 million worldwide. A lot of people watched it either in a theatre or streamed it in their homes. And while it is exciting to watch it's also one of the movies often noted for its scientific inaccuracies.

The plot of the movie is simple. NASA discovers that a large asteroid is going to hit the Earth. In response the space agency sends a rag-tag team of drilling experts to destroy the rock putting us in peril. The movie depicts the team traveling a great distance from Earth using a NASA space shuttle. They of course spectacularly accomplish their mission but in a way which has little or no scientific credibility.

What does this have to do with astronomy outreach? The movie is an example of how people can be unintentionally misinformed about science.

I frequently encounter misinformed people as a volunteer at the California Science Center in Los Angeles. On many Saturdays, I volunteer in the center's very popular *Samuel Oschin Pavilion* where the space shuttle Endeavour is on display. Over 20 million guests have visited Endeavour since it's opening in 2012. It is one of the most popular tourist-stops in the Los Angeles area. My role as a volunteer is to answer questions and encourage guests to explore and learn from the many exhibits.

An all-too-common question I am asked is "How many times did Endeavour fly to the Moon?" Those of us with more than just a casual understanding of spacecraft design understand that the space shuttles were built only for low Earth orbit (LEO). They were not engineered to go beyond the Earth as was depicted in *Armageddon*.

In the next building over, the California Science Center has on display three space capsules – a Mercury, Gemini, and Apollo capsule – all with signs explaining their role and the progression of America's race to the Moon. The Apollo capsule on display is a sibling of the eight capsules that did travel to the Moon. So why the confusion about which spacecraft traveled to the Moon?

To understand how people can make this mistake we must acknowledge how prevalent misinformation is in our culture and its effect on the unsuspecting public. Movies are one source of misinformation. And *Armageddon* is not the only movie that incorrectly depicted space shuttles. Since the 70's, the movies *Moonraker* (United-Artists, 1979), *Lifeforce* (Sony/Tristar, 1985), *SpaceCamp* (Twentieth Century Fox, 1986), *Space Cowboys* (Warner Bros., 2000), and *Gravity* (Warner Bros., 2013) all showed the iconic NASA space shuttle and its astronauts performing un-realistic tasks. All but one of these movies was rated PG or PG-13 so a lot of adults and children viewed them. The producers of these movies would no doubt say that they make movies for entertainment and not education – which is obviously true. But does a non-scientifically trained audience interpret it that way? Or are they led to believe that the movie is realistic or partially realistic? Seeing is believing right?

Most of us understand that we should suspend belief when we go to the movies. But are we unintentionally tricking our brains into becoming misinformed? It's a fair question to ask. One can argue that collectively these six movies may have misled the public into believing something that is not true.

Outreach Astronomers will no doubt meet people with similar misconceptions. During astronomy events I have been asked:

> Will Mars soon appear as large as the Moon?
> Were the Moon landings faked?
> Where is the planet Nibiru? Is it going to collide with Earth?
> I read that the planets are in alignment. Is this dangerous?

All the above are hoaxes easily found on the internet. And they regularly appear on social media sites.

It may be very easy to dismiss people who believe in crazy ideas, but while they may be outliers, they could also be leading indicators of the public's current misunderstanding of science. Questions like these can be the proverbial canary in the coal mine. The public's response to the 2020–2022 worldwide pandemic, including all the hoaxes and misinformation, has been identified by many public health experts as an indication of the public's lack of understanding and general trust in science.

In the absence of accurate information, misinformation can take hold of people's imagination and become assimilated as fact. The outreach astronomer plays an essential role combating misinformation and promoting science. Doing so may be very important for all of us.

1.3   Professionalizing Informal Education

It is well understood that formal education is structured; it is carefully planned, vetted, and assessed for effectiveness. Formal educators receive professional training on techniques and theory. Textbooks go through rigorous editing and review processes. Lesson plans developed by teachers are carefully designed. There can be no doubt that formal education is a profession.

Can the same be said about *informal educators*? Without training and carefully developed materials, informal education can be haphazard and ineffectual. At best this can lead to missing important opportunities to inform the public. At worst it can lead to misinforming the public as badly as fictional movies.

To address this issue, several higher education institutions offer degrees in informal education and museum studies. Professional societies have been formed with the specific goal of professionalizing the field by providing training and resources.

The next chapter lists some of these organizations and many of the best practices and techniques that they promote. Also listed are materials that are offered to help the Outreach Astronomer be the most effective educator possible.

Chapter 2
The Science of Outreach

Methodologies used in *Formal Education* are not an exact science, but are based upon the results of decades of research in the field of psychology. Like all sciences, however, its practices improve over time. Research conducted by psychologists over the past 200 years has shaped how educators teach children today. Understanding how people learn and knowing how to help them is key to being an effective Outreach Astronomer.

2.1 The Scientific Foundation of Education

While studying education in undergraduate and graduate school, teacher- candidates study various theories that explain how children learn. This field of study, known as cognitive development, was started by Swiss psychologist Jean Piaget in 1936. At that time Piaget theorized that children naturally construct and use mental models of the world, which he called "schemas." He proposed that learning occurs when the child tries to assimilate newly acquired information. If the new facts don't match an existing schema, then the child is forced to modify or create a new schema to accommodate the data. This organizational process is what we call learning.

Schemas can be thought of as patterns. Eating at a restaurant, for example, is the pattern of sitting at a table, reading a menu, ordering food, eating, paying for your meal, and tipping the wait staff. It is a very different pattern than eating at home.

Adults know this pattern and understand how eating at a restaurant works. But it would not be obvious to someone who has never eaten at a restaurant before. There are also variants of the pattern found in different cultures and at different types of restaurants. In some countries tipping is not expected, while in the United States not tipping is thought to be impolite. Fast food restaurants require payment upfront. At a fancy "white tablecloth" restaurant the bill is typically presented at the end of the

© The Author(s), under exclusive license to Springer Nature Switzerland AG 2024
R. Stember, *Share the Universe*, The Patrick Moore Practical Astronomy Series,
https://doi.org/10.1007/978-3-031-53495-9_2

meal. A child must learn the appropriate patterns before they can become a restaurant customer.

Outreach Astronomers can apply the concept of schemas and help children and adults construct their understanding of astronomy and science. Consider, for example, observing the planet Jupiter through a telescope. Its moons form a pattern that is similar to the structure of the solar system. It was observing this very pattern, through his newly acquired telescope, that Galileo Galilei became convinced that Copernicus' heliocentric model of the solar system was correct. Over the course of several days, Galileo observed that objects were orbiting around the planet. These observations supported his notion that the Sun is at the center of the solar system, and not the Earth as was generally accepted in the year 1609 CE.

Chapter 8 includes a hands-on activity for students to replicate what Galileo drew. By observing Jupiter and comparing drawings created over the course of an evening a class can see how the moons orbit around the planet. The students can see for themselves how Galileo came to believe in the heliocentric model of the solar system (Figs. 2.1 and 2.2).

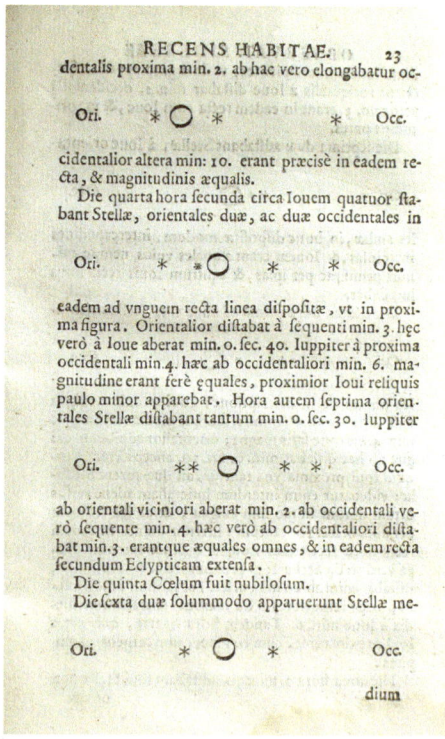

Fig. 2.1 Publication of Galileo's discovery of Jupiter's moons (Photo Credit: History of Science Collections, University of Oklahoma Libraries, CC BY 1.0 via Wikimedia Commons)

Fig. 2.2 A seventeenth
century telescope. (Photo
credit R. Stember)

2.2   Modern Education Techniques

The techniques employed by teachers have evolved greatly since the days of the one room schoolhouse and lectures from the chalk board. Modern educators have developed teaching techniques that better engage students and address various styles of learning.

It is widely understood that not all students learn the same way. Some children may do very well in a lecture style environment while others require a more hands-on approach.

When participating in events at schools you may come to recognize some of these teaching techniques. The way students ask questions, the type of questions asked, data that they are seeking, or the way they work in teams may offer clues. Being aware of these techniques can help you respond appropriately and enhance the benefit of the event for students.

Professional educators refer to "styles of learning." Some research suggests the existence of visual, aural, verbal, and kinesthetic styles. Other research contests this view and suggests alternative ways to think about the learning process. One of these alternatives is called the "Multiple Intelligences" theory.

Howard Gardner, in his book "Intelligence Reframed; Multiple Intelligences for the 21st Century" suggests that instead of multiple styles and one type of intelligence – people may exhibit multiple types of intelligences. Gardner identifies several intelligences, each of which may represent a strength or weakness in the individual's ability to learn:

Linguistic – effective use of language, spoken and written.
Logical-Mathematical – logical problem solving.
Musical – being able to perform and appreciate musical patterns.
Spatial – recognizing and using patterns in space.
Kinesthetic – associated with bodily coordination and movement.
Interpersonal – understanding moods, temperaments, desires and working with other people.
Naturalist – appreciating differences in environments and living things.
Existential – understanding "who we are", "what is love", etc.
Pedagogical – how to convey and share knowledge with other people.

I describe these competing ideas and some of the techniques employed not to suggest that the outreach astronomer should become a professional formal educator. Nor am I suggesting that you study these concepts and ideas in detail. These ideas are described so you can better appreciate what challenges educators face in the classroom. Not only do they have a specific syllabus to teach (more on that in Chap. 3) but they must do so effectively and appropriately for several classrooms a day, each with 30 or more children. And every child in the teacher's care can exhibit different intellectual strengths and weaknesses.

Most educators employ a classic *Direct Instruction* technique where concepts and skills are conveyed directly to the students. This technique is commonly associated with a lecture style presentation, but it can be combined with other techniques. Some of these techniques are described in the Fig. 2.3. The "What to Expect" column in the table lists what clues you might observe during an outreach suggesting which technique is being employed.

Knowing the instructions provided to the students before an outreach event, and the technique being employed, can help you make the event more educational for the participants. In general, it's always wise to:

1. Encourage students to ask the questions that *they* think are important.
2. Be willing to accommodate requests for specific information.
3. Recognize and work with teams of students.
4. Treat all students equitably.
5. Participate in the games and tasks assigned.

The Outreach Astronomer should also be flexible and ready to diverge from their original plan. This approach will be most helpful to the teachers and students and will enhance the student's learning experience.

TECHNIQUE NAME (GRADES)	DESCRIPTION	WHAT TO EXPECT
Flipped Classroom (grades 5-7)	Students study the material on their own time. Class-time is reserved for hands-on activities.	Questions, forms that students need to complete.
Kinesthetic Learning (all grades)	Learning through hands-on activities and demonstrations.	Hands-on activities that align with the subject matter.
Inquiry Based Learning (grade 7+)	Prioritizes students' curiosity and ability to analyze information.	Questions. Requests for specific data.
Project Based Learning (grades 1-12)	Students are taught basic concepts then tasked with solving real-world problems and issues.	Questions about specific topics and examples.
Problem Based Learning (all grades)	Students investigate problems first then are taught basic concepts.	Generalized questions followed by drill down into specific topics.
Collaborative Learning (grades 1+)	Students work as teams.	Equal participation. Interaction between students.
Thinking Based Learning (grades 1+)	Students are taught how to use critical thinking skills and creatively solve problems.	Specific questions, seeking explanations of concepts, terms, definitions.
Game Based Learning (all grades)	Employs games to teach concepts.	Completion stamps, specific activities recorded on a form or game board.

Fig. 2.3   Modern teaching techniques

The teaching techniques listed are best left in the hands of professional formal educators. They have learned and practiced how to best employ these techniques. But the educator will appreciate your willingness to coordinate and accommodate their efforts. It helps both them and you achieve your shared goal of educating students.

2.3   Informal Education Techniques

Informal Education is also not a science, but like *Formal Education*, it has progressed and benefited from techniques that have been developed and honed over many years.

To understand how *Informal Education* has progressed it is important to first appreciate the differences between the audiences of the two types of education.

2.3.1   Captive vs. Non-captive Audiences

It is widely understood that K-12 students are legally and socially required to attend school. Students generally understand this as do their parents. Additionally, most students know that if they want to get good grades, they must pay attention in class. The teacher will be testing them on the material presented. Paying attention, doing homework, and studying for tests are necessary to get at least a passing grade.

Informal educators call this type of audience a *Captive Audience*. Members of a captive audience are obliged to be there, and they act accordingly. For example, under normal circumstances, a student would not suddenly stand up and leave school without permission. Nor would they routinely skip their classes if they want good grades. They know both behaviors will result in a negative outcome.

The motivation of a member of a *Captive Audience* reflects these obligations. Even if they hate the subject matter, students understand that they are required to attend and participate. Furthermore, they are motivated to overcome any animosity that they may have to the subject matter, the teacher, or the environment.

Informal Education involves a very different type of audience – a *Non-Captive* audience. Understanding the nature and motivation of a *Non-Captive Audience* is key to understanding *Informal Education*. The members of a *Non-Captive Audience* chose to participate, not because they have too, but because they want to. They know they have the option to leave or "tune out" if the topic or the presenter is not interesting. There are no requirements to attend, nor any tests given. Getting a good grade is not part of the calculus.

It is this difference in the nature of these two audiences that makes formal education and informal education two very different endeavors.

2.3.2   When Informal Education Occurs

Unlike formal education, informal education can happen anywhere and at any time. It can be intentional (self-actuated) or accidental (passive). Individuals may seek out opportunities to learn or they may accidentally encounter and assimilate knowledge at unexpected times and in unplanned situations.

Most of a person's life does not involve a planned learning situation. Causal conversations with neighbors, watching a TV program, and visiting a store may be planned but not for the purpose of learning. None-the-less such encounters often do result in learning. For example, during a routine visit with your doctor, you may learn that a vaccine is now available for shingles. Not knowing how a person catches the virus you ask questions and learn that anyone who was previously infected with chickenpox, even as a child, carries the varicella zoster virus and can get shingles later in life. Your doctor conveyed information and you learned something new. You probably did not plan to learn something new during the visit other than getting good news that you are healthy.

An astronomy outreach event is also a type of planned event. All activities at the outreach are planned. The *Outreach Astronomer* planned to be there and, if it's at a school, the school staff also planned to be there. Attendees of the event likewise planned to be there.

But planning to be at an event does not necessarily mean that the attendees are seeking learning opportunities. A parent at a school's Astronomy Night might only be in attendance because their child needed transportation. The parents' presence does not mean that they are there to learn. Fortunately, as has been my experience, parents can often learn during a school event just like their children. The presence of the parents is an opportunity of which the Outreach Astronomer can take advantage. School events are not just for kids.

Likewise, a child at the event may be more interested in socializing than learning. School outreach events generate a lot of excitement, sometimes for secondary reasons. Regardless of the planning involved and the motivation for attending, the Outreach Astronomer should recognize that education can occur at any time.

2.3.3 *The Profession of Informal Educator*

The importance of Informal Education has been studied and recognized for at least a hundred years. There have for a long time been professionals working in the field such as museum curators, exhibit designers, planetarium directors, and educational outreach coordinators.

Recently the field has experienced a renaissance. Visit a museum or science center today and the evidence is easy to spot. Exhibits, films, display signs, and staff all may demonstrate how seriously their institution takes its role as a source of informal education. Today's museums are not the dust collecting exhibits of the past. Experts in the field work very hard to convey important and complex concepts to an interested lay audience. The way exhibits are designed, signs printed, videos and audios produced, have all been impacted by recent research into informal education.

This same passion for inspiring the public should be embraced by the Outreach Astronomer. Fortunately, several organizations have answered the call to help.

Funded by the Association of Science and Technology Centers and the National Science Foundation, the *Center for Advancement of Informal Science Education (CAISE)* provides resources to informal educators. Its objective is to support STEM education and advance the development of new learning tools.

The *National Science Teaching Association (NSTA)* offers information on outreach partnerships, lesson plans, books, and journal articles about informal STEM education. And the *National Informal STEM Education Network (NISE)* is a community of practice bringing people, opportunities, and resources together to further the impact of informal education.

The National Aeronautics and Space Administration (NASA) has created a group specifically for informal educators and museums. Called the NASA Museum and Informal Education (MIE) Alliance, this group makes available NASA images, video, and provides free training. Anyone involved in informal education can apply for membership.

NASA's Jet Propulsion Laboratory (JPL) also hosts the Night Sky Network for astronomy clubs and informal educators. This network provides many useful resources for the Outreach Astronomer.

Role Models for the Informal Educator

There are also several professions that can serve as role models for the Outreach Astronomer. Two well-known ones are the informal educators at museums and science centers; the other the *Informal Interpreters* employed at national, state, and local parks around the country.

Professional informal educators at museums, science centers, planetariums, and libraries take their educational roles very seriously. Specific academic degrees and extensive experience are often required to fill these positions. The description for a recent on-line job posting for a "Museum Program Specialist" serves as a good example: "...develop educational programs for schools, adults, and general audiences."

The posting indicated applicants should possess at least a bachelor's degree in a STEM related subject and a thorough knowledge of curriculum development.

Museums and science centers routinely hire experts in informal education for exhibit design, curation, acquisition, archival, and education.

The professionals who design exhibits are creating opportunities for learning at the institutions. Education department staff at these museums are specialists who know how to develop hands-on learning activities. These are just some of the professionals who make it possible for visitors to learn during a fun visit to a museum.

2.3.4   The Great Naturalists and the Technique of Informal Interpretation

Another role model for the Outreach Astronomer can be found in the tradition of the well-known naturalists and preservationists of the nineteenth and twentieth centuries. John Muir (1838–1914) and Freeman Tilden (1883–1980) both popularized and educated the public about the natural wonders of the United States. As they explained in their own words, they gave voice to the valleys, mountains, meadows, rivers, and plains which would eventually become our great national parks. They inspired people to learn and preserve those precious resources. Muir and Tilden laid the groundwork for what is known today as *Informal Interpretation*.

John Muir, who wrote about the Sierra Nevada Mountains of California, was the first to define interpretation. He wrote in his Yosemite notebook:

> I'll interpret the rocks, learn the language of flood, storm and the avalanche. I'll acquaint myself with the glacers and wild gardens, and get as near the heart of the world as I can.

Freeman Tilden would later redefine interpretation as connecting objects in the natural world with human meaning in experiential ways. He explained the difference between interpretation and simply sharing facts.

The work of these two men, and others, evolved into the modern practice of *Informal Interpretation*. This teaching technique is so useful it has been integrated into the training provided to the employees and volunteers at many park systems around the world including the *U.S. National Park System* (NPS). The NPS, under the program name *Interpretative Development*, trains many of its 20,000 employees and volunteers in Informal Interpretation. Nearly every NPS employee and volunteer who interacts with park patrons is given an opportunity to take this training.

According to the National Park Service:

> Informal interpretation requires knowledge of the resource (information), knowledge of the audience (customer service), and appropriate interpretive technique (recognition of the moment to tell a story, demonstrate an activity, or move from the tangible resource to an intangible story or message).

Educating park patrons about indigenous plants, animals, and geologic features at a national park is a lot like interpreting the night sky. Instead of floods, storms, and avalanches – Outreach Astronomers interpret nebulas. Instead of gardens and geologic features we share views of galaxies, stellar nurseries, and gravitationally locked star clusters. What we share exposes the universe to the public. And interpretation is what we do for our guests every time we share the eyepiece of our telescope. The ethos of the Outreach Astronomer is very much like that of John Muir and Freeman Tilden. We help others understand the wonders of our universe, its relevance to them, and the importance of preserving the night sky.

At some point during our life, we may all have been captivated by an informal interpreter. The occasion may have been a lecture, or a performance; it could have been a tour of a city, or local restaurants. You may have encountered this person while visiting a museum or a national park. I'm referring to events that were led by someone whom you had never met before; but somehow, this person captured your attention and imagination. You came away from the encounter feeling entertained and better informed.

Chances are you can remember specific attributes about the speaker. Maybe they were funny and entertaining; they had an air of authority; or were very insightful. And most likely they explained the topic in a way that you could easily understand and remember.

What this speaker did was likely carefully planned even though at the time it may have seemed improvised. It probably was not spontaneous or ad-hoc. Museums, parks, and tour group operators all work very hard to provide a quality product. And communication is an important ingredient. In this sense almost every phrase, if not word, from the speaker's mouth was planned in advance.

The leader of the event you remember was likely using the technique of *Informal Interpretation*. They knew how to make the interaction natural and conversational.

They used humor to entertain and keep your attention. Most importantly they made the subject understandable and relatable. You came away from the encounter with a deeper understanding because of how the speaker communicated. This is informal interpretation at its best.

2.3.5   The Elements of Informal Interpretation

Applying the techniques of Informal Interpretation can help the Outreach Astronomer better communicate and educate. Learning and practicing the technique will help you communicate with a non-captive audience in a manner that engages and keeps their interest. It will also help you organize your ideas so that the audience will find it easier to understand and assimilate what you share.

Informal Interpretation has several very specific qualities which make it different than other forms of communication. Fortunately, it relies on common techniques, many of which you already know. First let's identify the elements of Informal Interpretation.

Element 1 – Entertainment

As previously described, the audience of an outreach event is typically non- captive and therefore free to "tune-out" and even physically leave. Therefore, the interpreter must communicate in a way that keeps the audience interested. This often involves employing an informal speaking tone, encouraging audience participation, and incorporating humor.

Story telling is a close cousin to interpretation and the same techniques that are used in good story telling can be very helpful. Using action verbs, visual metaphors, exaggeration, and contrived situations often helps make a point and keep your audience engaged.

Action verbs convey power and capture people's attention. Verbs are vehicles that convey excitement and action. For example, saying "Comet Shoemaker- Levy 9 *crashed* into Jupiter" is far more interesting than saying "The planet Jupiter was *impacted* by the comet Shoemaker-Levy 9." The two statements convey the same information but the first is more exciting because it employs an action verb – *crashed*; the second statement by comparison is dry and boring.

Think of action verbs as conveying something that a person would do. You *crashed* your car as opposed to you *impacted* it. Personifying phrases in this way makes action real and meaningful.

Employing visual metaphors is another technique easily done in astronomy. A chart showing the comparably small size of Earth versus Jupiter conveys how massive Jupiter really is; three Earths can fit inside the Great Red Spot! A small almost hard to find circle next to a drawing of the Sun conveys that the Sun accounts for more than 99% of all the mass of the solar system; the small circle being the rest.

Incorporating humor is also very useful but it can also be a bit more challenging. Well-known actors Gregory Peck and Jack Lemmon both reportedly said "Dying is easy. Comedy is hard." Some people are better than others at being humorous. It's always best to stay within your own comfort zone and remember that not everyone appreciates the same jokes.

I typically use humor in my presentations but keep it low key in the beginning. Sometimes I will insert a joke just to gauge what my audience likes. If I get more than just a couple of chuckles, then I may push the envelope a little further. Even corny jokes can help keep the audience's attention. Interpretation is about conveying knowledge; you are not expected to be a standup comedian.

Asking your audience questions invites participation and keeps interest high. How much larger is Jupiter than the Earth? What is the nearest star? How many moons does Saturn have? These are some of the questions that I occasionally pull out of my "bag of tricks".

Always remember to be gentle in your response. An abrupt "wrong!" is a quick way to lose an audience. Instead saying "Good try. That's a common misperception people have" is a kinder way to respond. There is also a fine line not to cross. Asking too many questions can make the presentation boring. Your audience may begin to think that the talk is too much work and tune out.

There are many ways to make a technical talk enjoyable. No doubt there are many more ways to make it boring. A successful interpreter employs and practices the techniques with which they are most comfortable and can effectively use.

Element 2 – Relevancy

Making a subject relevant means connecting the subject with something that the audience already knows, understands, or cares about. Kids, for example, love everything having to do with explosions, dinosaurs, robots, and black holes. Describing a super nova using expressive terms and gestures is a sure way to get their attention. Relating the age of young stars to when dinosaurs roamed the Earth can help make the topic memorable.

Adults often connect to history or facts about current events. Describing how in 1054 CE, Chinese and Japanese astronomers reported seeing a star suddenly brighten in the sky is a great introduction to the Crab Nebula. It can help bring the subject matter home and opens the door to a discussion of how astronomers, both amateur and professional, are even today actively looking for Super Novas.

Element 3 – Organization

Organization is important to successful interpretation. Your presentation must be organized in a way that your audience finds easy to follow. Every fact and concept should build upon the previous one. Your listeners should never have to work very hard to understand what's being shared. This was emphasized by communications pioneer Dr. Wilbur Schramm. In his book titled "The Process and Effects of Mass Communications", Schramm explained the likelihood that a listener will assimilate new knowledge is inversely proportional to how much work is involved. He proposed the simple formula:

$$\text{Probability of Learning} = \frac{\text{Reward} - \text{Punishment}}{\text{Work Involved}}$$

The formula suggests that making a subject hard to learn, the denominator in the equation (work involved), overwhelms the nominator (reward – punishment). Schramm attributes the decline of movie theaters to the introduction of televisions into family homes. It's a lot less work to turn on the TV than get in the car, drive to a theater, buy tickets, and put up with crowds of people. Consider how many times you have made the same choice and opted to stream a movie at home instead of watching it on the big screen.

Organizing your presentation is a lot like playing with a set of building blocks. Every edifice requires a solid foundation. Erecting an arch on a flimsy base will fail. In the case of informal interpretation, the base is often what the audience already knows. They understand that stars produce light but may not understand the relationship of distance and luminosity. You explain that not all stars shine with the same luminosity. Some are brighter than others but also the further away they are the dimmer they appear. Concepts are building blocks placed one upon the another. Once luminosity is understood – then a discussion of the HR diagram and the different classes of stars can follow.

Organization also applies to the outreach event as-a-whole. The objectives of the event should be well defined and the communication that occurs should reflect these objectives. Outreach events typically occur within a finite period – typically a couple of hours. Objectives of short-term events should be modest such as raising awareness of specific astronomical phenomenon (ex. an upcoming solar eclipse) or helping students understand the structure of our solar system. Organizing communications so that they align with what students are learning in school, as described in Chap. 3, can help guide your communications.

Element 4 – Themes

Good interpreters present a theme as the main point. Think of a theme as a hook on a wall. Over the course of a presentation, you will hang related facts on the hook. Without the hook, the facts just fall to the ground and will be forgotten. But a clearly defined theme holds all the facts together and is remembered.

As an Outreach Astronomer I often share images of star forming regions in my telescope. The theme I present is "Stars form from nebulas." One of my favorite stellar nurseries is M42 – the Orion Nebula. It's a bright northern hemisphere nebula easily seen in small telescopes and binoculars during the winter months when schools are in session. This emission nebula can be found halfway down the sword hanging from Orion's three-star belt. It's even visible with the naked eye making it an easy target for a green laser. Pointing out a naked eye object, with a green laser, can help make the topic relevant and memorable to the viewer.

M42 is an active star forming region in our galaxy. The center area, named the *Trapezium*, contains young stars easily seen with even a small aperture telescope. These stars formed around one million years ago. That's what I want my guests to remember. I make it relevant, organize it in an easily understood narrative, and insert some humor:

> *This is a stellar nursery. Those are baby stars. Aren't they cute? They're only a million years old!*

The technique of *Informal Interpretation* can also be used in discussions about how science works (more on that in Chap. 3).

> *Astronomers have several theories to explain how stars form. How does the mass of a nebula coalesce into stars? Gravity alone? Is an external event needed? A super nova maybe? How does gravity overcome the radiative pressure that builds up internally? Astronomers have not solved this puzzle yet. But that's the fun of science. There is always plenty of research to be done.*

2.3.6 Astronomy Outreach as Informal Interpretation

All the elements of informal interpretation can be employed at astronomy outreach events. And every aspect of an event can benefit from the technique.

The theme of an outreach event itself should be well defined. It may be broad like "What's in the sky tonight?"; or it could be a narrow: "How lunar eclipses occur". The topic and objective of the outreach can be dependent on the composition of the audience or what the students are currently studying in class.

Signs displayed at the event should also incorporate themes. This is frequently evident at museums. The title of the informational panel is the theme. The rest of the copy fills in the details that relate to the theme. Printed materials produced for your outreaches should also be themed based.

The value of entertainment during the event should never be underestimated. Seeking entertainment may be the main motivation for people who attend an event. Outreaches are often advertised as "fun family events", so make it fun. If the outreach is not enjoyable, then not many people will stay. And if they don't stay, they won't learn.

Relevancy is often dictated by the age, experience, and interests of the audience. Your attempt to relate a topic to a third grader will be far different than what you do for the child's parent. The recipients' educational level and age can make some topics irrelevant. Chapter 3 discusses the astronomy related concepts that are taught to K-12 students. It will guide you on the topics that are relevant for students at different grade levels.

The Elevator Speech

At an outreach event you sometimes have precious little time with a guest during which to communicate a theme and supporting facts. I have seen many long lines of people waiting to look through a telescope. Sometimes guests get only a few seconds or a minute or two at the eyepiece.

Sales professionals use the term "elevator speeches" to describe the very short presentations that they give. The term comes from an imaginary scenario in which the CEO of a company joins the salesperson on a short elevator ride. The salesperson has only a few seconds to get the CEO's attention, make a sales pitch, and crack open the door to future sales. Salespeople often practice their elevator pitches over and over again, to be proficient.

Be prepared like a salesperson. Have your "elevator pitches" ready to be used during your event. If you are sharing a view of Jupiter, be prepared with a short speech about Jupiter, its great red spot, and its moons.

> We are looking at Jupiter now. It's the largest planet in our solar system. Ten Earths would fit across the face of Jupiter. Three Earths could fit in the Great Red Spot. If Jupiter was a glass bowl, more than 1,000 Earths would be needed to fill it.

Asking Questions and Active Listening

The techniques of *Informal Interpretation* can help an Outreach Astronomer have a larger impact. Remember the four elements of interpretation and apply them regularly to your efforts.

It's also important to understand the importance of asking questions and actively listening to your guests. Asking questions can help you determine what your guests know and what interests them. It involves your guests in the conversation and encourages them to be open to what you are sharing.

Active listening – hearing, understanding, and acknowledging what your guest says – is key to connecting with your guest. It shows respect and empathy. Learning is not a one-way street. It requires interaction and communication between two parties.

Asking a guest if they have looked through a telescope before will help you better help them use your telescope. Enquiring what they know about Jupiter can lead to a more meaningful discussion about the planet. Not knowing an answer is a welcome opportunity to learn; We are all learning and none of us know all the answers.

Education since the time of Socrates has emphasized asking questions. The famous Socratic method was a technique he pioneered that encouraged students to question what they know – or think they know. In the context of an astronomy event, it can help your guests apply schemas that they developed from a young age; or build new ones to accommodate new facts and concepts.

Asking questions and together discussing answers helps your guest formulate a solution. Most educators agree that it is better for a student to arrive at a solution than being told the answer. In the process of arriving, the individual is building the framework of understanding – the paradigm needed to explain the concept and facts.

Answering with Concepts

How you answer a guest's questions is equally important. Answers should provide context.

Q – How far away is the Sun?
A – On average the Sun is 93 million miles away from the Earth. If you could drive there at 65 miles per hour the trip would take 163 years.

Q – How long would it take to travel to another star?
A – The next nearest star is Proxima Centauri. It is 4.24 light-years away. If we sent astronauts there traveling at 35,000 miles per hour, which by the way is 10,000 miles per hour faster than the Apollo capsules, it would still take 81,000 years to get to there.

Including the reasoning for an answer is also important. The rationale helps the listener understand the concepts supporting the answer.

A – How do we know that our Sun will become a red giant in 6.5 billion years?
Q – Astronomers have studied and classified stars for hundreds of years. By categorizing them by their mass-luminosity relationship we have a good understanding of the life cycles of stars. Our sun is classified as a G2 star – a relatively low mass star that is fusing hydrogen into helium. Scientists have estimated the rate of that conversion and when the hydrogen will run out. When there is not enough hydrogen left to counter its own internal gravity, our Sun will become a red giant. It will repeatedly expand and collapse; each time giving up much of its mass into space. Eventually it will become a white dwarf and shine very dimly.

2.3.7   The Role of the Outreach Astronomer

It has been said that it is better to be "the guide on the side rather than the sage on the stage." Outreach astronomers can become better at what they do by using the techniques described in this chapter. They can also help the public understand a rapidly changing scientific landscape.

Imagine if you knew very little about astronomy and came across a newspaper article touting the discovery of the oldest galaxy ever detected. The article says that the galaxy is 13.4 billion years old and has the largest redshift ever measured. A new multibillion dollar space telescope made the discovery possible. What could you comprehend from this news without understanding the current estimated age of the Universe; the terms "Light Years" and "redshift"; or the concept of how distances to objects, which we have not traveled to, are measured? It may all seem incomprehensible. You may wonder if it is science or science fiction. How could you possibly understand the significance of the discovery without a strong foundation in astronomy? And why would you think that the money spent on this telescope was worth it?

The above discovery was in fact recently announced by NASA. The James Webb Space Telescope (JWST) and the discoveries that it is supporting are regularly covered in national and local news. We can look forward to possibly decades of new discoveries and advances in our understanding of the Universe because of the JWST.

Extrapolate the above scenario into other scientific fields and similar news coverage: climate change (2023 has been reported as the hottest year on record), evolution (recently a previously unknown humanoid species, Homo Longi, was discovered), stem cell research (biologists are learning how to reprogram cells), vaccine development (mRNA vaccines are revolutionizing immunology). Is it any wonder that there are so many people skeptical about and willing to reject science?

It is reasonable to surmise that public skepticism is at least partially rooted in a lack of understanding and appreciation for how science works. The Pew Research organization recently reported that only 51% of the U.S. public has a "fair amount" of confidence in scientists.

Based upon news reporting and social media posts, it is also safe to assume that the public is frequently confused, and maybe skeptical, about the objective of science. A few media outlets make it a point to publish articles implying that some research is a waste of taxpayer money. Clicks and ratings are often more important than taking the time to explain how science progresses and why the research is important. Rarely is it explained why professional organizations like the NSF and NIH found merit in the proposals and approved the grant funding. The public may not understand why it's important to study the "sex lives" of frogs but biologists and ecologists may. It could offer important information critical to our survival in a changing world.

Building an appreciation for science requires a concerted effort to better explain what scientists do beyond the simplistic model of the scientific method. This matters because people questioning science are likely to be voters. They elect leaders who decide on funding for scientific research. Leaders make policy decisions and pass laws some of which should be informed by science.

Most importantly, everyone deserves an equal opportunity to learn and benefit from knowledge regardless of their socio-economic status, race, religion, politics, gender, sexual orientation, strengths, and weaknesses. We all deserve the opportunity to learn and grow and are entitled to the benefits that modern science brings. Knowing how mRNA vaccines work and came about could go a long way to dispelling fears about this remarkable advance 30 years in the making. It could also save a lot of lives.

Astronomy is perhaps the most approachable of all the sciences. It captures the imaginations of both children and adults. Maybe because of the sizes and distances involved, or the powerful and violent nature of the Universe – astronomy fascinates people of all ages. Children love talking about black holes and Super Nova. Adults are often spellbound contemplating the vast distances and the possibility of finding life elsewhere. The Outreach Astronomer is well suited to engage with the public on these topics and the very nature of science itself.

The approachability of the subject can be used to strengthen the public's understanding of science in general. What is needed is a person willing to explain complex concepts in the simplest terms possible; someone, who with humor, compassion, and respect, is willing to accept people where they are on their intellectual journey;

someone willing to answer questions and explain the significance of new discoveries; someone who can place themselves "in the shoes" of a guest who is eager to learn.

It is a noble and rewarding calling – unequalled and unparalleled in many ways. You have in your hands the ability and opportunity to change people's lives in important ways.

Outreach Astronomer
Mr. Zachary Schierl – Astronomy & Physical Science Instructor, former NPS Park Ranger.

Zach took his first class in informal interpretation as a summer volunteer at Bryce Canyon National Park. After completing his graduate school studies, he accepted a year-round ranger position at Cedar Breaks National Monument where he hosted weekly star parties and dark sky interpretation programs for the public. His efforts included creating a "Master Astronomer Program" to teach local residents astronomy and how to advocate for dark sky preservation. His 40-hour workshop combined astronomy basics with informal interpretation training.

Zach's goal was to get the local community to appreciate the importance of maintaining dark night skies.

"I tried to make astronomy programs go beyond astronomy to capture the ecological, scenic, and economical value of dark night skies and the importance of controlling light pollution." – Zach Schierl

Zach's efforts paid off. It helped Cedar Breaks achieve International Dark Sky status in 2017.

Zach believes that the power of interpretation comes from the focus on relevance. He understands that a skilled interpreter is needed to help guests get beyond jargon and appreciate what they see through a telescope. This can be challenging to achieve during a short interaction, but he believes interpretation is the best way to help guests understand why astronomy is important and how it is relevant to them. As he says it answers the "So what?" question.

Currently Zach teaches astronomy and physical science at Yakima Valley College in Washington state.

Pictures reprinted with permission of Zachary Schierl.

Chapter 3
Know Your Audience

Understanding your audience is a prerequisite for a successful outreach event. The make-up of any audience can vary greatly. Generally, you would expect that adults know more about astronomy than do children; high schoolers know more than do middle schoolers, and so on. But the facts may be surprising.

This chapter discusses what we can expect from different types of audiences at outreach events and what studies show about their base of knowledge. With this understanding, the Outreach Astronomer can appropriately tailor their themes and messages.

3.1   What Adults Know

When conducting an outreach event, you rarely know in advance what your adult audience knows about astronomy. I have encountered both people with an advanced understanding of the Universe as well as others with very limited knowledge. In 1950 the United States Congress established the National Science Foundation (NSF) and its National Science Board. The act mandated that every year the board report on the state of science and engineering in the U.S. It's annual *Science and Engineering Indicators* report covers many topics including the public's perception and understanding of science.

Disappointingly, in 2002 the board reported that "… only 50 percent of Americans know how long it takes Earth to circle the sun… Only 21 percent of those surveyed were able to explain what it means to study something scientifically… and only a third knew how an experiment is conducted."

It's hard to imagine that half of Americans do not understand that it takes the Earth 365 days to orbit the Sun. Two-thirds do not understand how science works. If these statistics are still true, they are indicators of a serious challenge for the Outreach Astronomer not to mention the individuals who lack this basic knowledge.

© The Author(s), under exclusive license to Springer Nature Switzerland AG 2024   23
R. Stember, *Share the Universe*, The Patrick Moore Practical Astronomy Series,
https://doi.org/10.1007/978-3-031-53495-9_3

3.1.1   What Is Science and How Does it Work

Science is more than just a body of knowledge; it's also the practice of how that knowledge was developed. Sharing facts does little to help an audience understand how the facts came to be known. And it doesn't help the listener understand the more advanced concepts that build upon the facts.

Most people learn about the "scientific method" in grade school and not much more about the process of science. Consequently, the public has assimilated a very simplistic understanding of the term "theory" and not much about how science works. Rather than guesses or suggestions, as is often mistakenly thought, theories are working models (paradigms) that explain accepted facts. Over time scientists refine and build upon these models with further experimentation. Occasionally theories are refuted and replaced. More commonly they provide a framework for the continuous advancement of scientific knowledge.

The public needs a better understanding of how theories are developed and the role that scientific communities play in sustaining consensus about these paradigms. A very solid explanation of this process is presented in Thomas Kuhn's seminal work "The Structure of Scientific Revolutions". In his essay, Kuhn describes the important role played by scientific communities. Kuhn writes that these communities are uniquely specialized and competent professional groups. They consist of researchers and educators in specific fields of study. From the outside we may vaguely know them as evolutionary biologists, or cosmologists, or astrophysicists, or computational neuroscientists, or any one of the seemingly endless and growing number of specialized branches of science.

By their nature, these communities are intellectually isolated from the public. Lay people are not likely to understand how these groups function nor the subject matter that they address. But these communities provide an efficient way to train new scientists and, most importantly, promote progress in their fields.

The paradigms accepted by these communities provide efficient ways to educate new members. Physics students for example need only study science textbooks and not read canonical texts such as Newton's *The Principia*. Hence, physics and other scientific fields can be taught in an efficient and expedient manner. Even so it can take many years beyond undergraduate school to achieve the recognition and status afforded to senior members of a scientific community.

Each scientific community is the recognized arbitrator of progress in their field. Generally, only members (e.g., people with the proper education and credentials) can propose changes to accepted paradigms and be taken seriously. Progress is incremental and focused on issues that are known about each paradigm. It may be a slow, even sluggish, process but this is a beneficial function of science. Science is not built to provide immediate answers. It is designed for building long lasting paradigms.

The public often expects science to provide answers on demand which is understandable during a pandemic but unrealistic when the goal is comprehension and durability. Often a news article will appear that describes a single scientific study that shows some benefit or risk associated with a medicine, food, or environmental factor. The public will grab hold of the results as if it's a new scientific principle. One experiment, of course, does not change a paradigm or a theory. The process takes many experiments, often over years, and even different types of studies to fully tease out all the nuances of the data. While the public does not have the patience for such rigor – scientific communities do.

Unlike the public's impression, research is typically performed by teams. Long gone are the days of the researcher working alone in the laboratory. Large pharmaceutical companies, for example, employ thousands of chemists – organizing them into departments, subdepartments, and teams – each working on a specific target, disease, or condition. Just one FDA approved medication can cost $1 – $2 billion and require dozens of years of R&D and clinical trials.

Scientists working in pure research are often tasked with solving known issues with existing scientific paradigms – "mopping up" as Kuhn calls it. This is often the realm of cosmologists and theoretical physicists. As this process progresses, proposed solutions to issues must be accepted by most of the related community before it is incorporated into the paradigm. Such is the purpose of scientific journals, peer reviewed articles, and scientific conferences.

All these processes are deliberately slow, rigorous, and laborious. It works well precisely because of these attributes. Over hundreds of years the scientific process has matured in such a way that it serves humankind very well. For proof, consider that the average life expectancy at the beginnings of modern science was around 35 years old. Today, life expectancy in first-world countries has more than doubled. This is largely due to advances in medicine and science.

All of this may be a mystery to the average lay person. But this structure and process is incredibly important to understand if one wants to understand science. Through rigorous experimentation and consensus, modern science progresses and the truth of how the universe works is unveiled.

In some ways this structure is like that employed in other professions such as law and medicine. Both have professional communities, aka state bars and medical boards, which are officially and legally, tasked with training, testing, and onboarding new members. Law students, for example, first must attend an accredited law school, then pass a bar exam. If they pass, they can become licensed to practice law in their state. As it should be, these communities are isolated from the public. The public gets no input on who can become an attorney or a physician. That's entirely up to the professional community. It's an official contract the public has with the professions; codified into law by the legislatures of all 50 states and territories of the United States of America.

Unlike professions in law and medicine, scientific communities have no official authority, other than the intellectual license they grant to themselves. Because the public has so few opportunities to interact with scientists directly, this isolation may have led to a reduction in the public's trust in science. As a result, the motive of scientists is sometimes questioned. Science itself has even been wrongly likened to a religion by some people. And long accepted scientific paradigms have been rejected by small but significant segments of the general population.

A concerted effort is needed if trust and confidence is to be sustained in science. The Outreach Astronomer can help build that trust by educating the public about how science works.

What Adults Know About Astronomy

It is not uncommon to encounter adults at events who do not understand the basics of astronomy. From planets to galaxies, from nebulas to stars – concepts taught in elementary and middle school may have been long forgotten or never learned.

For this reason, I often start a conversation at an outreach with some basic definitions. For example, while showing M42 – the Orion nebula – in my telescope, I may tell my guest:

> We are looking at a star forming region in our galaxy called the Orion nebula. A nebula is a large molecular cloud in space. Like many other objects in the sky, we measure distance with what we call 'light-years.' Light travels at a fixed but extremely fast speed. This nebula is so far away, that even as fast as light travels, it still takes 1350 years for the light to reach us here on Earth. That means that we are seeing what this nebula looked like 1350 years ago.

With this explanation I have:

- Defined a nebula (a molecular cloud in space).
- And introduced the concept of light years as a unit of measurement.

The purpose of this part of my narration is to help gauge what my guest knows. I don't want to leave them behind by going deeper into topics if they don't understand the basics. I encourage my guests to guide me in the conversation by asking questions.

As described in Chap. 2, I follow the basic structure of informal interpretation. I state the theme (stars form from nebulas) and keep my guests entertained with facts that are organized and relevant. As described in Chap. 2, facts don't just hang out there; they must be part of a story that is related to the theme (hung on the hook).

Once the basics and terminology are established, the presentation can turn to related facts:

> Stars are forming from this nebula. We can see four stars in the center area which is called the Trapezium. These stars were formed primarily from the hydrogen that is present in the nebula. We believe that they are less than a 1 million years old.

The above narration builds upon what the guest knows; or, as Piaget theorized, the schema that they use to explain stars. We are adding to that schema and expanding the guest's knowledge. Any topic can be approached in this manner to achieve the objectives of your presentation and outreach event.

3.2   What Is Taught in School?

When doing an outreach event for school age children it's best to understand what they are learning in school. Before an event I often ask the teachers, what topics are currently being covered or will soon be taught. Aligning your activities with the teacher's lesson plans can be a great benefit for the students. It can "turn on light bulbs" that otherwise may be currently off.

Most STEM teachers in the U.S. create lesson plans based upon the *Framework for K-12 Science Education* from the National Research Council (NRC) or the Next Generation Science Standards (NGSS) that are derived from it.

3.2.1   The NRC Framework for K-12 Science Education

A significant step to improve K-12 science education in the U.S. was accomplished in 2012 when the National Academy of Science and the National Research Council (NRC) published its *Framework for K-12 Science Education*. The framework was based upon research done in the mid to late 1990s. It identified what was lacking in science education up to that point and what needed to be improved. The goal of the NRC was to enumerate the core science and engineering concepts, ideas, and thinking processes that 12th grade students need to possess by the time that they graduate. The objective was to identify for teachers what is needed for their students to evaluate and consume information, continue education, and develop skills for careers (STEM related or not).

Shortly after the release of the framework, a consortium representing 26 U.S. states was formed by the National Science Teachers Association (NSTA), the American Association for the Advancement of Science (AAAS), and the National Research Council. The goal was to put the NRC's framework and performance expectations into practice. While appearing to be complicated and a large departure from past techniques, the result has been described as a holistic way to teach science. The consortium's work product became known as the *Next Generation Science Standards* (NGSS). This standard lays the groundwork for how K-12 science will be taught in U.S. schools possibly over the next 20 years or longer.

Since its release in April of 2013, 20 states and the District of Columbia have adopted the NGSS. Another 24 states have developed their own standards based upon the NRC *framework*. According to the National Science Teachers Association (NSTA) this covers approximately 71% of all U.S. students as illustrated in Fig. 3.1.

The Framework for K-12 Science Education identifies what every student should know, the techniques and skills they should possess, and how they can think in ways that are applicable across many disciplines. It provides a multi- dimensional scheme as listed below (with examples):

State Adoption of NGSS and Framework

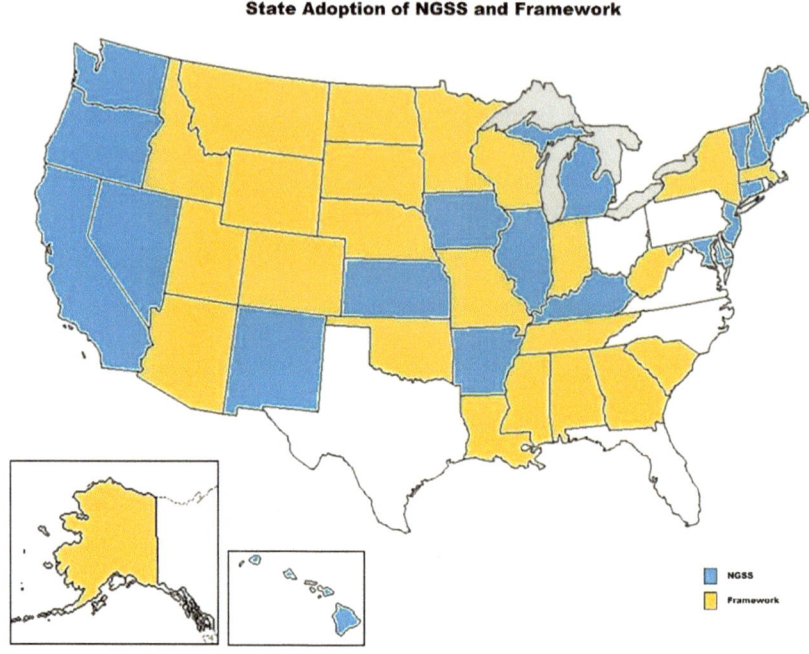

Fig. 3.1 States that have adopted the NRC Framework and NGSS

Science and Engineering Practices – (ask questions, make and use models, conduct investigations).
Disciplinary Core Ideas – (atoms & molecules, mass & gravity, electromagnetism).
Cross Cutting Concepts – (looking for patterns, cause & effect, structure & function).

The framework categories instruction into four domains of study:

Physical Science (PS)
Life Science (LS)
Earth and Space Science (ESS)
Engineering, Technology, and Application of Science (ETS)

Earth and Space Science (ESS) includes astronomy concepts identified as ESS1A and B in the framework. Each of these sections are repeated in subsequent years (Grades 2, 5, 8, and 12) with progressively more detail and nuance. The end points of each section are listed on the following pages by grade and topic. The end points are what students are expected to know by the completion of the school year.

ESS1A – The Universe and its Stars

Grade 2-	Patterns of motion of the sun, moon, and stars can be determined and described.
	At night many stars can be seen with the naked eye.
	A telescope makes it possible to see many more stars and observe the moon and planets in detail.
Grade 5-	Our sun is a star. It is closer than other stars. As such it appears to be brighter than other stars.
	Stars vary in size and distance from the Earth.
Grade 8-	Models can be used to describe the patterns of motion of the sun, moon, and stars.
	The Universe started with the Big Bang – a period of rapid expansion. Our galaxy, called the Milky Way, is just one of many in the Universe.
Grade 12-	Our sun is expected to have a life span of around 10 billion years.
	There are more than 200 billion stars in the Milky Way, and the Universe consists of hundreds of billions of galaxies.
	Spectroscopy is a technique that is used to identify the composition of stars, their movement, and distance from Earth.

ESS1B – Earth and the Solar System

Grade 2-	Sunrise and sunset vary by season.
Grade 5-	We can observe patterns caused by the orbit of the Earth around the sun, and the moon around the Earth. This includes day and night, differences in the length of daylight during the seasons, phases of the moon, different positions of the sun, moon and stars by day, month, and year.
	Some objects are naked eye objects while others require a telescope to see. Planets are not always visible from the Earth.
	Constellations are patterns of stars and have been used for navigation. They move across the sky because of the rotation of the Earth.
Grade 8-	Gravity keeps the planets and asteroids of our solar system in orbit around the sun. Moons, including our own moon, orbit around their respective planet.
	This model also explains why the Earth experiences tides, eclipses, and the apparent motion of planets in respect to stars.
	Seasons on Earth are the result of the tilt of the Earths axis. This results in differences in the intensity of sunlight reaching different parts of the Earth.
Grade 12-	Kepler described the elliptical motion of orbiting objects. Orbits can change by collisions and the gravitational effects of other bodies.
	Long term cyclical changes in Earth's orbit have resulted in ice ages and climatic changes.

Other educational end points intersect with astronomy and may be considered by the outreach astronomer when crafting communications. These include:

PS1.A Structure and Properties of Matter

Grade 2-	Different types of matter exist in either solid or liquid form.
Grade 5-	Matter can be divided into particles too small to see but the effects of them can be detected, felt, or seen.
	The amount of matter as indicated by weight is conserved when it changes form.
Grade 8-	There are approximately 100 different types of atoms which make up matter. Molecules form by the combination of differing numbers of atoms. Pure substances represent a single type of atom or molecule and have distinct physical and chemical properties.
	Gases and liquids are molecules or inert atoms that are in motion. In gases, the atoms or molecules are further apart than in liquids. In solids, atoms are close together and may vibrate but do not move in relationship to each other.
	Temperature and pressure can change the state of matter.
Grade 12-	Atoms have a nucleus made of protons (+), neutrons (no charge), surrounded by electrons (−).
	The periodic table organizes elements by the number of protons (horizontally) and similar chemical properties (vertically).
	Interactions of atoms is determined by electrical forces.

PS1.B Chemical Reactions

Grade 2-	Heating or cooling may cause observable changes. Some changes are reversable (melting) some are not (burning).
Grade 5-	Mixing two or more substances can create a new substance with different properties. Total weight of the substances does not change (the difference between mass and weight is not taught at this grade level).
Grade 8-	How substances react can be predicted. The resultant molecules have different\ properties than the constituent atoms.
	The number of atoms does not change, and mass is conserved. Some chemical reactions release energy. Others store energy.
Grade 12-	Chemical reactions and their rates are the result of molecular collisions and rearrangement of atoms and changes in total binding energy.
	Reactions and properties of matter explain many important biological and physical phenomena.

PS1.C Nuclear Processes (Grades 2 & 5- topic not covered)

Grade 8-	Nuclear fusion is the merging of two nuclei (hydrogen) into a larger nucleus (helium) and results in the release of energy.
	Fusion releases more energy per atom than any chemical process and requires extremely high temperatures and pressures such as found in the core of a star.
	Fusion in a star releases light and produces more massive atoms. Supernovas also produce heavier elements.
Grade 12-	Fusion, fission, and radio-active decay are the result of changes in binding energies.
	Total number of protons and neutrons does not change in a nuclear process.
	Strong and weak interactions determine stability and the processes involved.
	Radio-active decay follows the exponential decay law and can be used to date materials by comparing isotope ratios.
	Average stars stop producing light when the end product of fusion is carbon. More massive elements are formed only in more extreme conditions provided by supernova.

PS2.A Forces and Motion

Grade 2-	Objects pull with each other when connected, Push on each other when colliding. Push and pulls can have different strengths and directions changing the speeds and directions of the objects involved.
	An object sliding across the surface of another object experiences friction.
Grade 5-	An object at rest can have multiple forces acting on it if they zero out each other. Forces can change an object's speed and/or direction.
Grade 8-	Newton's Third law – Whenever one object exerts a force on another object, the second object exerts an equal and opposite force on the first.
	The motion of an object is the result of the sum of all the forces acting on it.
	The larger the mass of an object, the greater the force needed to achieve the same change in motion.
	Describing positions and directions of objects requires identifying the reference frame and units.
Grade 12-	Newton's Second law – The acceleration of an object depends on the mass of the object and the amount of force applied. This law does not apply at the subatomic level without modifications or for objects traveling close to the speed of light. (Quantum physics and relativity are not covered at this grade).
	Momentum is mass times velocity. Momentum is conserved in any system.

PS2.B Types of Interactions

Grade 2-	Objects in contact can push on one another and cause a change in motion or shape.
Grade 5-	Objects in contact exert forces such as friction, pushes and pulls.
	Gravity, magnetic, and electric forces can affect objects that are not in contact.
	The strength of the above forces depends on the distance between and properties of the objects.
	Gravity pulls objects toward the center of the source of gravity.
Grade 8-	Electric and magnetic forces can either attract or repel. The strength of these forces will depend on the magnitude of the electric charges, current, or magnet.
	Gravity always attracts. All objects exert gravity.
	Gravity, electric, and magnetic forces can extend through space as fields. Gravity over large distances governs motions of solar system objects and galaxies.
Grade 12-	Newton's law of universal gravitation – objects attract other objects proportionally to their mass and inversely proportional to their distance.
	Coulomb's law states that electrical attraction or repulsion is inversely proportional to the square of the distance.
	On the atomic scale electric attraction and repulsion explain the structure, properties, contact forces, and transformation of matter.
	Gravity, electric, and magnetic forces can extend through space as fields. Gravity over large distances governs the motions of solar system objects and galaxies.
	Strong and weak nuclear forces control what happens inside atomic nuclei.

PS2.C Stability and Instability in Physical Systems

Grade 2-	Objects stay still or move depending on forces pushing or pulling it.
Grade 5-	Changes in systems can be directional and cyclical depending on the forces acting on it. Systems can appear to be static if multiple counteracting forces are applied.
Grade 8-	Stable systems can go to other states with even small forces. Repeating cycles imply patterns can be predicted.
	Some systems incorporate feedback mechanisms to maintain stability within limits.
Grade 12-	Systems can change in predicable and unpredictable ways. The greater the number of system constituents the harder it is to predict.

PS3.A Definitions of Energy (Grade 2- topic not covered)

Grade 5-	Objects that are moving faster contain more energy than slower objects.
	Energy can be transmitted via moving objects, sounds, light, and electric currents.
Grade 8-	The energy inherent in a moving object is called kinetic energy. The amount of kinetic energy is proportional to the mass of the object and increases with the square of its speed.

	Objects and fields can store energy dependent on their spatial positions – called potential energy.
	Thermal energy is caused by the motion of atoms and molecules. Heat is the transfer of thermal energy between objects or systems.
	Temperature is a measurement of the average kinetic energy of particles of matter.
Grade 12-	The energy of a system can be quantitatively measured and is dependent on motion and interactions of matter and/or radiation.
	A system's total energy is conserved as it is transferred between objects and forms.
	Mechanical energy is a combination of motion and stored energy.
	Chemical energy refers to energy released or stored by chemical reactions.
	Electrical energy is energy stored in a battery or transmitted by electrical currents.

PS3.B Conservation of Energy and Energy Transfer

Grade 2-	The Earth is warmed by the Sun.
Grade 5-	Moving objects, sound, light and heat indicate that energy is present.
	Colliding objects transfer energy and can release energy in the form of heat and sound.
	Light and electric currents can also transfer energy.
Grade 8-	Changes in an object's motion indicate a change in energy. Friction can increase thermal energy in both the object and the surface it is on.
	Making an object move requires input of energy.
	Conduction, convection, and radiation is the transfer of energy.
Grade 12-	Energy cannot be created or destroyed. The energy total for a system always equals the energy transferred into or out of the system – called the conservation of energy.
	Systems always trend to stable states.

PS3.C Relationship Between Energy and Forces

Grade 2-	To make an object move faster a bigger pull or push is required.
Grade 5-	Energy is transferred between colliding objects and can cause changes in motion.
	Magnets can exert force from a distance on other magnets and magnetized material. This causes transfer of energy between the two objects.
Grade 8-	Energy is transferred between interacting objects. The gravitational energy of an object is increased when an object is raised, and the energy is released when the object falls.
Grade 12-	Gravitational, electric and magnetic fields can transmit energy from one object to another. When the objects change relative positions, the energy stored in the field between them changes. Moving away reduces the stored energy.

PS3.D Energy in Chemical Processes and Everyday Life

Grade 2-	Objects rubbing against each other produces friction, warming them.
	Friction can be reduced in different ways.
Grade 5-	Producing energy involves converting stored energy into another form such as generating electricity by releasing water behind a dam. Food and fuel are stored forms of energy. Plants capture light energy from the Sun.
	There are ways to concentrate and store energy.
Grade 8-	Sunlight, carbon dioxide, and water are required for plants to create food (such as sugar) in a chemical reaction. (Photosynthesis is not taught at this grade).
	Burning fuels and digesting food requires oxygen in a chemical reaction to release energy.
	Reducing friction in a machine can improve its efficiency. It also reduces energy transfer such as heat to the parts of the machine and the environment.
Grade 12-	Fusion in the core of the sun releases energy in the form of radiation. Photosynthesis captures solar energy in a complex chemical process. Man-made solar cells convert solar radiation into electricity.
	Living organizations transport and transfer energy via several different physical and chemical processes.

PS4.A Wave Properties

Grade 2-	Waves can be made in water. Waves cause some of the water to rise and some to lower – but the water does not move in the direction of the wave.
	Sound can make objects vibrate and vibrating objects can make sounds.
Grade 5-	Waves can differ in amplitude and wavelength. Waves can be additive or can cancel each other out depending on their phase.
	Earthquakes are seismic waves traveling through the Earth's crust.
Grade 8-	Waves can have a repeating pattern with an identifiable wavelength, frequency, and amplitude.
	Sound waves need matter to travel through.
	Waves, such as seismic waves, can reflect at boundaries.
Grade 12-	The boundary between two different media, and the properties of the media, affects how waves reflect, refract, or are transmitted.
	Information can be encoded in waves of different frequencies, digitized, and stored.

PS4.B Electromagnetic Radiation

Grade 2-	Objects are visible only when light is either illuminating them or the object is giving off light. Very hot objects can give off light.
	Light can pass through some material. Other materials completely block or allow some of the light to pass through them.

	Objects that block light produce shadows.
	Mirrors and prisms can change the direction of light.
Grade 5-	Stars and the sun produce a lot of light some of which reaches the Earth.
	Objects become visible when light reflects off its surface and enters our eyes. The color(s) we see depends on the object and the color of the light source.
	Lenses can magnify images of objects by bending the light.
Grade 8-	Light that hits an object is either reflected, absorbed, or transmitted depending on the properties of the composition of the object and the color of the light.
	Light normally travels in a straight line except at boundaries between different transparent materials. Lenses and prisms are examples of this effect.
	Light can travel through space where there is no matter. Scientists describe light both in terms of waves and as particles.
Grade 12-	Electromagnetic radiation can be thought of as both waves of electric and magnetic fields and as particles called photons.
	Quantum theory unites these two models (no further explanation of the theory is provided at this grade).
	Objects like atoms cannot be observed with waves that are larger than the object.
	All forms of electromagnetic radiation travel through a vacuum at the same speed (the speed of light).
	Light and longer electromagnetic radiation is absorbed by matter and converted to thermal energy.
	Shorter wavelengths, such as UV, X-rays, and Gamma rays, can ionize atoms. This can cause damage to living cells (for example sun burns and cancer).
	Elements absorb light at specific frequencies which can be useful for identification.

PS4.C Information Technologies and Instrumentation

Grade 2-	People's senses inform them about the world around them. With our eyes we can see light. With ears we can hear sounds. And with touch we can feel vibrations.
	We also use devices to communicate over distances.
Grade 5-	Eyeglasses, telescopes and microscopes incorporate lenses so we can see more. Lenses bend light.
	Information can be digitized (such as pixels of an image) and stored. Computers and cell phones can also convert information from one form to another.
Grade 8-	There are many types of devices that can be used to detect signals.
	Digitized information can be transmitted as pulses of waves.
Grade 12-	Modern technology relies on our understanding of waves and matter.
	Scientific research depends on modern technologies.
	Quantum physics made it possible to manufacture semiconductors, computer chips, and lasers. (Quantum physics is not taught at this grade).

Summarizing the NRC

As the previous pages show, students by the end of the second grade have been introduced to the concept of stars, and the motions of our Sun and Moon. They come to understand that the times for sunset and sunrise vary by season. And that telescopes help us see distant stars.

By the end of grade 5, students understand that the Sun is a star that is closer than the others. Stars vary in size and distance. The orbit of the Earth around the Sun, and the Moon around the Earth, causes patterns that we can observe – seasons and lunar phases. They learn about the solar system, and they come to understand that lenses bend light.

Eighth graders learn about the Big Bang theory and that the Milky Way is just one of many galaxies in the universe. Gravity is described and the model of the solar system is used to explain eclipses of the sun and moon. They learn about fusion taking place in the center of stars.

It's not until the twelve grade that students learn about absorption of light at different wavelengths by atoms – the beginning of understanding spectroscopy. They are introduced to quantum physics, and technologies that are used for scientific research.

Remembering these endpoints can help the outreach astronomer plan for an event and better communicate with students so they understand the themes and facts presented during an event.

3.3   Other Audiences

Outreach events are often offered to specific groups outside the context of school. Each audience may present important challenges or requirements for successful and relevant presentations.

3.3.1   Girls Scouts of the USA

The Girl Scouts was founded in 1912 and has become the largest U.S. organizations specifically serving girls. Membership is organized by school grade:

Level	Grades
Daisy	K–1
Brownie	2–3
Junior	4–5
Cadette	6–8
Senior	9–10
Ambassador	11–12

Like other organizations, the Girl Scouts offers advancement opportunities based upon merit and educational achievement. Literacy badges are offered at each of the grade levels on various topics including astronomy. New Space Science Badges were introduced in 2018/19. These badges include an observing component with which an Outreach Astronomer can assist.

Daisy's (grades K-1), being the youngest, are engaged using playful forms of discovery. For example, Daisy's may be asked to spot the first star that appears in the evening sky. They are asked to observe naked eye objects such as the Moon and easily located star clusters like the Pleiades. The Outreach Astronomer will find that a green laser is very useful to point out objects and illustrate stories about the constellations.

The next older group, Brownies (grades 2–3) are old enough to look through a telescope. Viewing the Moon, Saturn, and Jupiter are favorite targets. The scouts can experiment with different eye pieces and discuss what differences they see. In school they are learning about seasons and changes in sunrise and sunset.

Juniors (grade 4–5) are exploring patterns of the solar system in school. They are ready to learn about the constellations' slow movement to the west over the year and how to use a planisphere. In school they may have already learned about lunar phases.

Being middle school students, Cadettes (grades 6–8) are learning about light and how it can be separated into colors. By the end of the eighth grade these students will have learned about gravity, planets, and asteroids. This, unfortunately, is also around the age that girls sometimes decide that science is not for them. Encouragement, supportive mentors, appropriate interactions, and role models can be very important at this age.

Seniors (grades 9–10) are learning more complex concepts about the universe, the life cycles of stars, and astronomy in general. The Outreach Astronomer can introduce them to the Hertzsprung-Russell (HR) diagram, the life cycles of stars, optics, and telescope design.

As the most mature group, Ambassadors (grade 11–12) may be ready to do citizen science projects and explore a future in STEM. NASA missions, and female astronauts offer inspiration, and the scouts may be ready to begin their own path in amateur astronomy. Understanding career paths and possibilities is important at this age.

At every one of these levels, outreach astronomers should be aware of the language used and the messages sent, both intentional and unintentional. Professional STEM fields are still largely male dominated. The Pew Research Center reported in 2021 that even though women comprise just over half the general workforce, they make up only 40% of the nation's physical scientists and 15% of the engineers. These figures have stubbornly remained constant and have changed only 1% since 2016.

Research indicates that boys and girls show no significant difference in abilities associated with STEM subjects. Why fewer girls ultimately go into STEM fields is thought to be influenced by lower confidence and negative self- perceptions. Outreach Astronomers can help counter this trend with encouragement, countering the idea that math and science is hard, and sharing information about female role models.

With funding from NASA and help from the SETI Institute, Girl Scouts USA created new space and astronomy badges in 2018/19. The new badges are designed to encourage curiosity, confidence, competency, and an understanding of the value of STEM subjects. Each activity highlights contributions made by female astronomers, both historical and current day figures. There are a great many female role models to share with girls and boys alike. Some of the female scientists mentioned in the activity booklets are:

Caroline Herschel – Discoverer of many comets; contributed to the first catalogs of nebula and star clusters; co-researcher with her husband William Herschel.

Annie Jump Cannon – Defined and classified over 225,000 stars by their spectra using the O, B, A, F, G, K, M temperature-based system currently in use.

Henritta Swan Leavitt – A deaf scientist who discovered the relationship between the brightening and dimming of Cepheid Variable stars and their absolute magnitude establishing a "standard candle" used today to measure distance.

Dr. Kimberly Ennico Smith – Astrophysicist who served as project scientist for the New Horizon's mission to Pluto and the SOPHIA infrared observatory.

Nathalie A. Cabrol – Planetary scientist and astrobiologist for the SETI Institute and director of the Carl Sagan Center; Developed NASA's strategies for the exploration of Mars.

Dr. Madhulika Guhathakurta – Astrophysicist who is the Lead Program Scientist for NASA's Living with a Star (LWS) program and co-chair of the international LWS Committee on Space Weather.

Sara Seager – Astrophysicist and planetary scientist studying exoplanets; professor at MIT.

Biases and Microaggressions

Sometimes subtle phrases and stereotypes are carelessly used in conversation. Verbal missteps can have unintended and subtle consequences. Not so many years ago it was common practice for parents to encourage boys to play with trucks and building sets while girls were given dolls and dresses. Today psychologists refer to these actions as biases and microaggressions.

Researchers uncovered these biases when they asked young girls and boys to draw a picture of a scientist. Frequently the picture drawn by the children was of a male and not a female. This stereotype was built by a constant flow of subtle messages the children had been receiving their whole lives. Girls did not think scientists could be women.

Generally, children may also be led to believe that either you have what it takes to do math and science, or you don't. This is a common belief even among adults today.

I was never good at math. I just don't get it. I'm better at other things.

This attitude is referred to by psychologists as a *Fixed Mindset*. It's the idea that you are born with an intelligence, and it is not something you can control. The truth is that anyone can learn a subject. No one was ever born knowing how to ride a bicycle nor how to decode DNA. It had to be learned.

The preferred concept is that intelligence is developed through effort. This is referred to as a *Growth Mindset*. It recognizes that failure is expected. How many times did we fall off the bicycle when we were learning to ride? Each time we fell, we got up and tried again until the skill was finally mastered. Similarly, learning math and science will involve many "falls." It's important that the learner understands that this is normal. Eventually we all can learn a subject if we just stick with it.

Using gender neutral language, appropriate examples, and encouraging a *Growth Mindset* can help engage girls and counter the idea that STEM is not in their future. Including female role models in your presentations can help level the STEM playing field.

NASA has made inclusive STEM engagement a high priority. Their *Women in STEM* page at https://www.nasa.gov/stem/womenstem.html offers many resources to help engage girls and show them that science is not just for boys.

The Jet Propulsion (JPL) *Night Sky Network* also provides links for learning how to specifically assist girl scouts earn badges. The page lists local scout troops (see bit.ly/astroall) that you can connect with.

The *Nightsky Network* web page provides links to videos about communicating with girls. These videos were produced by the *Astronomical Society of the Pacific* (www.astrosociety.org). They offer tips for communicating in an inclusive and constructive fashion and specifically address:

- **Unintended Biases and Micro-messages** – words and actions that, while maybe unintentional, can be perceived as being negative, unwelcoming, and counterproductive.
- **Growth Mindset vs. Fixed Mindset** – encouraging the idea that anyone can learn science. Too many people assume that they are just not good at science. Failure to understand need not be feared or avoided. It's trial and error that leads to growth.
- **Representation and Role Models** – Historically fewer women enter STEM fields than men. Fortunately, this trend is slowly changing. Equity can be achieved by sharing representative role models.

There are literally hundreds of female role models that we can refer to in our outreach efforts. It's a matter of making the effort to learn about and incorporating their stories into what we say. The *NASA Women in STEM* web page is a great resource for learning and sharing the stories of these women scientists, engineers, technicians, flight controllers, and astronauts. What could be more inspiring for a young person than seeing someone like themselves, someone with a similar background, or herstory, achieve what they dream of becoming.

In 2022, the *Girl Scouts Research Institute* (GSRI) reported that significant progress was being made in building positive attitudes about STEM. In a quasi- longitudinal study of 1255 girls, GSRI compared the attitudes of middle school and high school scouts who participated in Girl Scout STEM activities to those who did not. It found that girls who participated in the activities were 20 percentage points more likely to be interested in pursuing a STEM career. By providing fun and supportive STEM activities, the Girl Scouts USA is making a real difference in encouraging more girls to learn about and pursue a STEM career.

3.3.2 Scouts BSA

Formally known as the Boy Scouts of America, Scouts BSA now welcomes both boys and girls. Scouts BSA is organized by age :

Level	Age
Cub scouts	5–10
Scouts	10–18
Venturing	14–21
Sea scouts	14–21
Exploring	16–20

Outreach astronomers can play an important role in helping both Cub Scouts and older scouts understand concepts in astronomy. Scouts work toward proficiency in many topics including science and engineering. Merit badges are offered for both Astronomy and Space Exploration. The requirements for all merit badges are listed at https://www.scouting.org/merit-badges.

To receive the astronomy merit badge, scouts need to be able to:

- Identify at least 10 constellations.
- Demonstrate or explain how binoculars and telescopes are used for astronomy.
- Identify eight major stars.
- Explain the motion of the planets.
- Describe the composition of the Sun.

Scouting BSA actively enlists the help of trained volunteers. Outreach astronomers interested in becoming a Merit Badge Counselor should contact their local council (see https://www.scouting.org/about/volunteer/).

3.3.3 Religious Groups

I have been asked many times to run outreach events at a church or for a religious organization. At first my inclination was to kindly decline out of fear of conflict with their belief structure. I didn't feel equipped, or comfortable, being put in a position where I would possibly have to argue about the age of the Universe, or about the Big Bang, or other scientific theories. And I didn't want in some way to appear to be questioning their beliefs.

After participating in a few of these events I decided that my fears may have been misplaced. Rather than being brought unwillingly into arguments, the participants were eager to explore much in the same way I did when I bought my first telescope. The heavens are an awesome sight regardless of your religious beliefs and finding commonality is not hard at the eyepiece.

It is worth remembering that many astronomers and scientists, from Galileo to Einstein, had and have strong religious beliefs. A significant number of modern-day scientists believe in some form of deity and have found ways to reconcile those beliefs with modern scientific paradigms.

Religion does not have to be in conflict with science. Scientific "truth" refers to understanding how the Universe works. It is not the same as the "truth" that religious orders seek. While I may not share the religious beliefs of my hosts, we are able to share in admiring the beauty and complexity of what we see. I find that I am able discuss astronomy while still respecting my guests' religious perspectives. And they often respect my perspectives.

Some suggestions that may help for these type of events:

- Familiarize yourself with the beliefs that the group generally holds, particularly about creation and the age of the Universe.
- Avoid subjects that are in conflict with those beliefs – unless you are prepared and motivated for a debate. If so, keep it friendly and respectful.
- Be respectful of differing beliefs. Insisting someone is wrong rarely alters minds.
- Be open to different interpretations and ideas. Science requires an open mind. Model that foundational aspect of science.
- Don't expect that every fact you share will be accepted.
- Focus on science as a method to understand how the Universe works.
- Explain that science only tries to answer questions that can be tested. Many questions cannot be tested and are out of the realm of science (such as the existence of a deity).
- Explain that science is self-correcting by design. Scientists don't expect to have all the answers nor that their solutions won't change. This is its strength and not a weakness.
- Discourage the thought by some that science is just another religion. This sets up unnecessary competition.
- Point out that many scientists hold strong religious beliefs. Conflict is not a given between science and religion because the goals of the two are very different. Many people find comfort and meaning in both at the same time.

3.3.4   Non-Western Cultural Groups

There are many cultural groups whose beliefs are not based on western ideas and science. Having been raised in the United States, and being from European stock, my education has been firmly rooted in the western cultural tradition. This is likely true for most people reading this book.

The belief structures and cultures of many groups in North America and elsewhere are not based on a western foundation. Native Americans, and indigenous peoples, across all 50 states and the 10 provinces of Canada, have their own belief structures and what they consider to be their "science."

Emphasizing commonality and respect opens the door to new learning opportunities. I have enjoyed learning about Native American culture and belief systems as they relate to astronomy. A wonderful source is the Indigenous Education Institute (http://indigenouseducation.org/). This institute offers a series of online lectures by renown native speakers describing their groups' understandings as compared to western science.

Native "science" broadly recognizes the centrality of relationships. The interaction and codependences of the environment, people, plants, and animals is central to their thought. While sometimes dismissed by western thinkers as mythology – indigenous stories and perspective can offer insights to better land management, and stewardship of the environment. It can inform modern society and maybe even western science.

Many ancient indigenous peoples were avid observers of astronomical phenomena. They labored over centuries to develop a technology of astronomical calendars to forecast the movement of animal herds, best times for planting and harvesting crops, and performing religious and political ceremonies.

Spread across North America are approximately 50 medicine wheels, some dating to before the construction of the great pyramids of Egypt. We believe that these "cairns" were used for observing astronomical phenomena. One of these wheels, the Majorville Cairn in Alberta Canada, is thought to be 4500 years old. Another, the Bighorn medicine wheel, outside of Sheridan Wyoming, is approximately 3000 years old. One of its spokes is oriented to the sunrise that marks the start of the summer solstice. An analysis of other cairns shows similar spokes pointing for example where the stars Aldebaran, Rigel, and Sirius rise at the start of the summer solstice, and consecutive periods of 28 days that follow.

The objectives of western astronomy and ancient indigenous science are different – but they do share an interest in observing the sky. Appreciating this common interest, and respecting native thought for its emphasis on relationships, can help open the door to exchanging knowledge generated by these two very different "ways of knowing."

When conversing with a Native American audience I find it helpful to:

• Acknowledge that I am a learner, an explorer, and wish to understand their perspective.
• I share what I believe to be true in friendship and with respect.

- Western science has the ability to build knowledge but recognizes that there is much more to understand.
- I believe that knowledge and wisdom are not the same. Science is about knowledge. The rest of society must learn how to develop that knowledge into wisdom.

Outreach Astronomers often share stories with their audiences from the ancient Greeks and Romans. The stories of the constellations Casseopeia, Andromeda, Hercules, Cephus and others are very entertaining. Why not also learn and share some of the stories from native peoples?

North and South American, African, Asian, Australian, and Oceania indigenous peoples all had tales that they told about the Sun, the Big Dipper, the Milky Way, Orion, Taurus, Cygnus, and the Pleiades. There is the Inuit story of Nanuk, a kind and generous polar bear that was being pursued by three hunters. These figures are represented by the constellation we know as Taurus. Another story, told by the Cherokee people, describes how a group of animals succeeded in moving the Sun from the dark side of the planet, where it was kept by greedy people, to bring light to the world.

One of my favorites is the story from ancient India of Svana the dog. It tells how the star Sirius came to be in the sky. The story goes like this:

> The beloved King Yudhisthira set off on his final journey to heaven high up in the Himalayan Mountains. He eventually succeeded in reaching the top while others who accompanied him fell ill and died. A homeless dog, he named Svana, strangely and unexpectantly accompanied him for his entire journey. When Yudhisthira reached the summit he met Indra the King of Heaven, who welcomed him but told him that the dog could not join him in heaven. Indra explained that the dog was the spirit of King Yudhisthira's father who wanted to make sure that his son safely completed his journey. King Yudhisthira entered heaven and Svana was rewarded with a permanent place in the sky as the star Sirius.

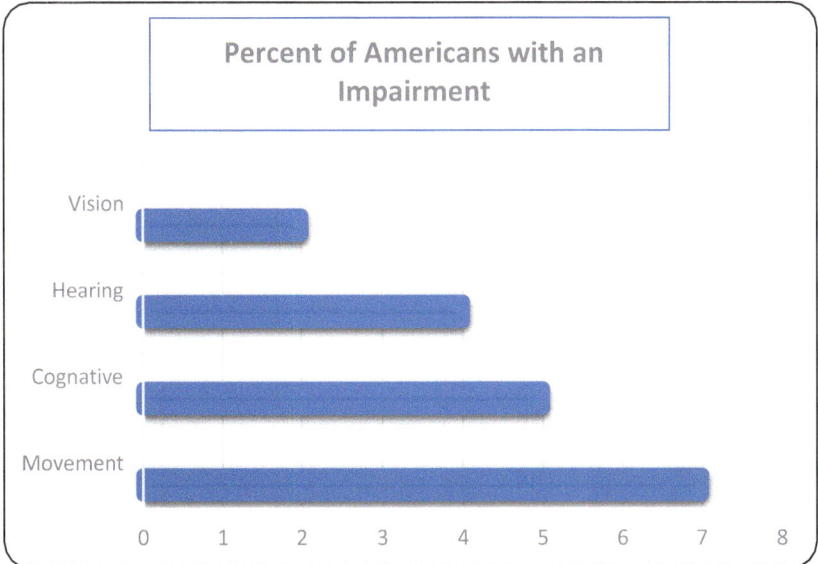

Fig. 3.2 Percentage of Americans with disabilities

Interestingly, Sirius is known as the dog star and is related to canines in several ancient cultures. It was also keenly observed by the ancient Egyptians as it would mark the annual flooding of the Nile River.

These are very entertaining stories to share but they can also give us insight into how people lived and the things that they valued. These legends, western culture included, were all efforts by people to better understand the Universe and themselves. This desire to understand hasn't changed for thousands of years. We should find ways to celebrate our efforts both ancient and modern. It's one of the things that makes us human.

3.3.5   Audiences with Disabilities

In 1990 the U.S. congress passed the landmark federal civil rights law known as the Americans with Disability Act (ADA). This important law provides protections against discrimination for Americans who have a disability. It requires that employers make reasonable accommodations for disabled employees and state, local, and public venues provide appropriate access to their facilities.

The ADA doesn't name every possible impairment covered by the act, but the U.S. Census bureau has defined several broad categories and estimated the number of Americans that have each disability as shown in Fig. 3.2.

With such significant percentages, it is highly likely that one or more guests at any outreach event has an impairment.

In 1997, the U.S. Congress revised earlier legislation and passed the Individuals with Disabilities Education Act (IDEA). Every five years this act is updated with new definitions, the last being done in 2017. The IDEA act requires the use of Individual Education Plans (IEPs) for special education students and Individualized Transition Plans (ITPs) to prepare them to enter society as an adult.

IDEA currently defines 13 categories of disabilities:

- Autism
- Deaf-blindness
- Deafness
- Emotional disturbance
- Hearing impairment
- Intellectual disability
- Multiple disabilities
- Orthopedic impairment
- Other health impairment
- Specific learning disability
- Speech or language impairment
- Traumatic brain injury
- Visual impairment including blindness

Providing specific strategies for each of these disabilities is beyond the scope of this book, but there are organizations that study and develop best practices for each. One of these organizations, Interstellar Inspirations, LLC (https://www.laurajeanchecki.com/), focuses specifically on formal and informal astronomy education. With funding provided by the National Science Foundation, this unique organization has developed materials specifically for many of the disabilities listed by the IDEA act.

3.3.6 Blind and Low Vision Audiences

People with blindness and low vision (B/LV) have been historically left out of astronomy events. This does not have to be the case. There are many tools and techniques available that can support full participation. Below are suggestions that you can incorporate into your event:

1. Fully describe your event in advertisements and web pages so that B/LV guests can plan their visit ahead of time. Instead of providing just a map, add text that fully describes the location. Text can be converted to audio by automated reading software used by some B/LV people.
2. Create a document that describes the layout of the event. For example:

   A table with materials describing Jupiter's moons and planned NASA missions will be located immediately to the right of the event entrance.

3. Low vision guests may need to closely examine materials. Use large print in handouts and signs.
4. Items which will be touched or picked up should be situated such that they can be easily put back in place. Use trays to prevent objects from rolling or falling off the table.
5. Have tactile reading materials on hand. Haptically Speaking publishes several tactile books on astronomy subjects (https://hapticallyspeaking.com/). Other materials and sources are listed in Chap. 5.
6. Be prepared to verbally and verbosely describe objects in the telescope. If possible, combine it with physical models and braille pictures.

While volunteering at the California Science Center I was once approached by a sighted individual who had a blind companion. They both were interested in learning more about the space shuttle Endeavour and asked if I could answer questions. Fortunately, I was prepared to do more than just answer a few questions. In my shoulder bag I had models to help.

I positioned myself directly in front of my two guests and introduced myself. I explained that I had a model of the shuttle and asked if they would like to hold it. With their permission I placed the model in my blind guest's hand and gave him a few seconds to feel the shape and hardness of the model. I proceeded to describe Endeavour in as vivid and descriptive language as possible:

The space shuttles were designed to launch into space on a rocket and land like a glider. So, Endeavour is shaped much like a jet liner. It's roughly the size of a Boeing 737. Most people think spaceships need to be made of hard metal. But the upper parts of the space shuttles are actually covered in soft white thermal blankets. I liken them to an oven mitt. The blankets are made of silica fabric and fiber glass, quilted together, and glued to the aluminum body of the shuttle. They are there to protect the shuttle from the tremendous heat generated during reentry. As the shuttle re- enters the atmosphere the air around it gets heated to almost 3000 degrees Fahrenheit.

The lower part of the shuttle – its belly – is covered with thick black tiles made of silica and coated with glass. All 24,400 tiles on Endeavour are glued to the body of the shuttle. They can withstand temperatures twice as hot as lava. No metal can survive that temperature, but silica can. The tiles are also very delicate. You could crumble one in your hands.

At that point I offered to take the model back and hand him an actual shuttle tile. I continued my narration:

Surprisingly lite, aren't they? The tiles are made mostly of air. It's a foam made of sand – or more accurately silica fibers. Silica is a wonderful insulator. Imagine going to the beach on a sunny hot day. If you remove your sandals your toes will feel the hot top layer of the beach. Now imagine digging your toes an inch or two down into the sand. What happens? Your toes feel cooler right? That's because sand – or silica – does not transfer heat very well. It's a very good insulator. That is why the space shuttle is coated in silica.

During my discussion I described the shuttle in vivid details so my guest could "see" it in his mind. I handed him physical objects to hold to help him "see" with his hands. For all I knew he could see some shapes and colors, many people who are legally blind have that ability. By his reactions, and follow up questions, I could tell that he was getting a full experience that otherwise he would not get.

I must have spent a full hour with these two guests. They asked many questions about the shuttle's design, purpose, and history. And it was all possible because I was prepared with a simple model and a tile. I was also ready to use language to help my guest "see".

Sighted people generally have many misconceptions about the blind. Getting past these stereotypes and misperceptions is a prerequisite for effective communication. My experience at the science center was very different from my first encounter with a blind person. At the age of 18, I had the pleasure of meeting Mr. Al Sperber, a radio show host and advocate for the blind. He was collaborating on a book with my parents at the time. That awkward first meeting opened my own eyes. While blind people face many the challenges – the greatest may be how they are treated by the sighted.

An astronomy event can offer a full experience for B/LV people. It takes practice and planning. But the effort is always appreciated by your guests. There are many specialized materials, braille books, and 3D models available, with which you can prepare yourself. Chapter 5 lists many of these materials and sources.

When working with a B/LV individual I find it helpful to:

- Position yourself directly in front of your guest so they know where you are located.
- Never hand them an object or touch them without permission. If granted they will extend the hand with which they want to hold an object.

- When leading a B/LV individual invite them to hold your elbow and if accepted, help them place the hand that they extend.
- Don't be embarrassed about using words like "see" or "look". Most B/LV individuals are accepting of this from sighted people and won't mind.
- Converse naturally but remember to use descriptive terms as much as possible.
- Dispose of stereotypes and misconceptions.
- Use words instead of facial expressions and hand motions. Instead of pointing at an object say *"Let's talk about [name of object] which is on your right hand side at 2 o'clock roughly three feet away."*

3.3.7 Deaf and Hard of Hearing Audiences

Hearing loss is a common issue for many individuals. Approximately 48 million Americans are dealing with some level of hearing loss.

Working with people who are deaf may be new to most readers. The Deaf Communication Hearing Center, a non-profit serving parts of Pennsylvania, New Jersey, and Delaware, offers very good tips:

- Ask the person how they prefer to communicate.
- Face the person when communicating.
- Maintain eye contact.
- Use normal, not exaggerated, facial expressions.
- If the person can lip read – speak at a normal pace.
- Use pictures and other visual aids when possible.
- Reduce background noise.
- Use technology such as a computer, cell phone, paper pad, or white board to communicate.
- Relax and be patient as the process may be new to you.

Addressing the needs of deaf or hearing-impaired people may also be accomplished by employing the services of an ASL (American Sign Language) interpreter.

When working with an ASL interpreter, ask what you can do to better support their interpretation. They may offer suggestions about your slides, handouts, or even how you give your presentation. The National Association of the Deaf publishes an on-line list of ASL interpreters at https://www.nad.org/deaf-interpreters-directory-list/.

3.4 Universal Instructional Design

While identifying best practices for all disabilities may be out of the scope of this book, Outreach Astronomers can make their presentations and event materials accessible to a wide range of participants by applying principles of what's called *Universal Instructional Design*.

Chapter 2 briefly summarizes basic educational theory – how people learn. *Instructional Theory* focuses on creating an environment in which learning can better occur. Allied with *Instructional Theory* is *Universal Instructional Design* (UID) which specifically addresses designing educational materials and environments such that they are accessible to a wide variety of people including those with disabilities. UID is an extension of the architectural accessibility concepts that were originally developed by Ronald Mace at the North Carolina State University.

Educators have adapted Mace's concepts and identified seven principles that define best practices for *Universal Instructional Design*.

Instructional materials and learning environments should be:

1. Equitable so a wide variety of people can use them.
2. Flexible to accommodate differing abilities.
3. Simple and intuitive to use.
4. Clear and useful in their instructions and communications.
5. Supportive and accommodating to the users.
6. Require a minimum amount of physical ability.
7. Allow sufficient physical space for use and access.

Consider the example of a physical activity a person with low vision would perform. This person would benefit if a large font size was used in all printed documents that are part of the activity. This addresses UID principles 1, 4 and 5.

A person with complete blindness on the other hand may be totally reliant on touch. Designing a surface with raised edges, like a tray, can inform them where the boundaries are for the activity. This addresses UID principles 1, 2, 3, 6 and 7.

To address principle numbers 4 and 5 volunteers assisting at the activity should be trained to verbally describe the lay out of the table:

> We have on the table in front of you three models of deep space objects. From left to right they are the Pillars of Creation, the Crab Nebula, and Eta Carinae. I suggest picking up the Pillars of Creation object first. It is on your left at 11 o'clock. Notice that this huge nebula has three molecular clouds extending out into space. It's at the edges of these structures that new stars are forming.

Activity volunteers can be trained to become the eyes for the blind and low vision participants. The descriptions and instructions they give should be concise and complete. Be generous in the use of positional information. Like any guest they should be given an appropriate amount time to explore and comprehend the information being shared. Ask questions and let them ask questions. They will guide you just as you guide them.

Guests who have issues with mobility or dexterity can benefit from appropriately sized objects and environments. Picking up a very small object may pose a real problem for people who have limited use of their fingers. Smooth objects may pose a problem for people who only use their palms. Adding texture or rough surfaces can help.

Leave a space at the table for a wheelchair. Position monitors and computers where everyone can see them regardless of their height. If steps or ladders are the only way to access a telescope – consider employing Electronically Assisted Astronomy as described in Chap. 5.

Applying these principles and techniques can take time and practice. And it may not be possible to create individual materials that address all disabilities. But over time your collection of objects and materials will grow to accommodate more and more people.

Testing materials with disabled volunteers can help improve materials and designs. This can be done at events or even before events using focus groups.

Flexibility is key for the outreach astronomer. Be ready to make modifications even on the fly if needed. Observe and learn just as your audience is learning and don't be afraid to ask for their advice.

Outreach Astronomer

Ms. Noreen Grice – Founder of *You Can Do Astronomy LLC*. Astronomy educator, DEI, and accessibility advocate.

After receiving an M.S. in astronomy from San Diego State University, Noreen returned to Boston to work at the Charles Hayden Planetarium. As a planetarium educator and coordinator, Noreen came to understand the need for new approaches to make astronomy accessible for blind audiences. She pioneered the printing of tactile pictures and booklets as supplemental materials for her planetarium shows. In 1990 the Boston Museum of Science published her first tactile book, *Touch the Stars*.

In 2004, Noreen founded her company *You Can Do Astronomy LLC* and now offers STEM workshops for teachers and students, as well as tactile exhibit design services. Noreen is the author of seven books on making astronomy accessible. She is the recipient of numerous awards recognizing her efforts from various organizations including the National Federation of the Blind, Boston University, the YWCA, and the Astronomical Society of the Pacific.

Noreen has several suggestions to help Outreach Astronomers work with blind audiences. This includes making plaster models of lunar craters and displaying them next to images of the craters (see picture above). She also recommends obtaining a copy of her book, *Touch the Stars,* which includes 19 tactile images including the Big Dipper, Lyra and the Summer Triangle, the Ring Nebula, Waxing and Waning of the Moon, Solar and Lunar Eclipses, Saturn, Jupiter, a globular cluster and more.

Pictures reprinted with permission of Noreen Grice.

Chapter 4
Planning and Running an Outreach Event

Many astronomy outreach events are simple star parties with telescopes set up for observing. These types of events may be stand alone or part of a larger event such as a school's annual campout. There are many other kinds of outreach events that can have very different setups, themes, objectives, and audiences. They include presentations, training classes, discussions, question & answer sessions, and much more. Regardless of the type and purpose of the event, thorough planning is always needed.

4.1 Basic Organization and Planning

While not often thought of as such, astronomy outreach events have many similarities to other types of public and private events. There are many details to consider, plan, organize, and monitor. During the planning stage of an astronomy outreach you may find it helpful asking the following questions:

- Who is the expected audience?
- What is the objective of the event?
- Where will the event be held?
- When does it take place (dates/times)?
- Who oversees the venue?
- What is your role and the roles of others who are involved?
- How many volunteers and support staff are needed?
- How will it be marketed and advertised?
- What is the budget?
- Will it be free to attend, or will there be an entrance fee charged?
- How will tickets be sold, and admission enforced?
- Does the venue require proof of insurance?
- What type and amount of insurance is required?

© The Author(s), under exclusive license to Springer Nature Switzerland AG 2024   51
R. Stember, *Share the Universe*, The Patrick Moore Practical Astronomy Series,
https://doi.org/10.1007/978-3-031-53495-9_4

The answers to these questions may be obvious for very simple events. But events that involve multiple people may be far more complex and require substantial planning. It is not uncommon for a school or public parks department to require the organization running the event to provide proof of insurance. Liability insurance and workers comp insurance may be required even if only one person is involved. Getting the required Certificate of Insurance (COI) from your insurance company can often take several days. Knowing that the proof of insurance is needed early on helps in the preparation for the event.

Schools have also been known to require background checks for the volunteers and staff who will be present at an event. In these cases, the school typically provides a form along and instructions explaining how and where to go to have the background check performed. There are several commercial companies that offer various levels of background checks. Their services range from on-line searches of public records, on up to rigorous investigations performed by state law enforcement and the Federal Bureau of Investigation (FBI). Different states and school districts will have different standards for the procedure.

Some states have enacted laws requiring that all organizations providing youth services perform background checks on board members and volunteers. The state of California recently enacted AB506 which requires background checks on all board members and volunteers who provide more than a certain threshold of labor each year.

Cities and municipal parks typically require that you apply for a special event permit weeks in advance. Abiding by city and park regulations are mandated. Simply advertising a star party and showing up at a park is often not allowed.

Figure 4.1 shows the event planning form Science Heads Inc. uses to capture and communicate details about a scheduled event. The form helps ensure that requirements are identified and addressed as early as possible in the planning process. As additional information becomes available it is added to the form and distributed to staff and volunteers. In this way everyone involved is informed and encouraged to understand their tasks and roles for the event.

For events involving many tasks and people, using project management forms and software can help keep the planning process moving forward. There are many cloud-based applications that can be used to keep track of tasks, who they were assigned to, and the status of each. In many cases simple spreadsheets can be created to manage the process from start to finish. For complicated projects and events, Science Heads Inc. uses the Task Planning form shown in Fig. 4.2.

4.2 Planning for Specific Audiences

Chapter 3 describes what you can expect with different audiences. Planning always should start with understanding your audience and their motivation. There are also other aspects of the audience that may need consideration in your planning.

OUTREACH EVENT PLAN

Title		Date(s)	
Location		Time(s)	
Host Contact(s)		Email(s)	
Objectives/Goals			
Audience		Expd/Act	
SH Activities			
Notes			

| # SH Vols Needed | | # Host Provided Vols | | EventPlanningForm.xlsx |
| Special Tasks Reqd | | | | |

PRE-EVENT PLANNING

	TASK	COMPLETED
a	Schedule Date/Time/ Venue	
b	Send out email requesting volunteers	
c	Send out confirmation email to Volunteers	
d	Send email to host confirming event	
e	Send email to volunteers with details of event	
f	Arrange/place advertising with:	
g		
h		

DAY OF EVENT

	Assigned To	Role/Task	Email/Telephone	Hours	Notes
1					
2					
3					
4					
5					
6					
7					
8					
9					
10					
11					
12					
13					
14					
15					
16					
17					
18					
19					
20					

Fig. 4.1 Example of an event planning form

EVENT PLANNING TASKS/STATUS

| Event: | | Location: | |
| Date of Event: | | Notes: | |

Task #	Task Desc	Assigned To	Due Date	Start Date	End Date	% Complete
1	Inspect site					
2	Request COI					
3	Send COI to host					
4	Schedule volunteers					
5	Pull kits from storage					
6	Distribute background ck forms					
7	Confirm background cks complete					
8	Send details to volunteers					
9	Check on venue status					
10	Prepare MOBS					
11	Check weather conditions					
12	Send last minute changes					

Fig. 4.2 Task plan for an event

4.2.1   Children

Very young children generally don't know how to look through a telescope. It's generally thought that children younger than 5 years of age won't be able to use a telescope successfully. They may not understand the concept of "looking through" the eyepiece, closing one eye, or how to maneuver their body such that their eye aligns with the visual axis of the eyepiece (see more about this in Chap. 5). Planning for and providing alternatives can help ensure participation and enjoyment by young children.

4.2.2   Aging Adults

As individuals age, climbing steps and ladders can not only be challenging but dangerous. Many amateur astronomers employ step stools and ladders alongside their telescopes. Because the eyepiece may be high off the ground not everyone may be able to look through the telescope and keep two feet on the ground. Step stools and ladders may be fine for children and younger adults – but can pose a safety concern for aging adults.

If very young children or aging adults are in the audience, alternatives to directly looking through a telescope, such as described in Chap. 5, can be offered. Consider using Electronically Assisted Astronomy and physical models instead of telescope viewing.

4.2.3 English as a Second Language

In the United States, many languages are spoken besides English. In communities where English is a second language (ESL) the Outreach Astronomer should be prepared by learning common phrases in the dominant language of the community. Knowing simple phrases and words in Spanish, for example, has helped me give directions to guests where I live in California. Safety is always of primary concern. Knowing simple words like "cuidado" (watch out – be careful), and "mirá" (look here) are useful.

Bilingual signs also can help ESL participants. When creating table tent signs and other display signs, include a translation provided with on-line tools such as Google Translate or Bing Translate. Both translate over a hundred different languages. While the translations may not be perfect, your participants will likely be able to understand the meaning and appreciate the effort.

4.3 Defining Objectives

While planning an event, close attention should be paid to the objective and goals of the event. If it's at a school and the hosting teacher has indicated she wants her students to see planets and moons, then this goal should be clearly shared with all the volunteers and staff. If the objective is to help a group of Brownies complete their Space Science Adventurer badge, then the goal may be to teach them how telescopes work and how to use one.

In both examples, the planning required doing some research - discussing the objectives and goals with the teacher; understanding the requirements of the Brownie Space Science Adventurer badge.

4.4 Selecting a Venue

The location where an event is to be held often requires compromises and accommodations. There are few perfect public places for an outreach event. The Outreach Astronomer typically must work with what's available. Some venues may offer a great dark sky but are so remote that few people will be able to attend. Other locations may offer plenty of parking and easy access but are surrounded with bright lights that can't be turned off. These are factors over which we have little control.

Once a venue is selected there are many things that are in our control and should be addressed.

If you have several sites to choose from, considering the amount of light pollution may be worthwhile. The darkness of geographical locations is quantified in several ways. One common measurement is the "Bortle Dark Sky Scale", named after John E. Bortle who created the nine-level scale. The Bortle Scale is a semi-quantitative measurement related to visual observations. For example, at a location rated a value of 1 (Excellent) Messier objects are visible using just the naked eye. Constellations become hard to decern because of the vast number of stars visible. The Milky Way illuminates the ground.

Another way to gauge darkness (and the lack of light pollution) is using a Unihedron Sky Quality Meter or software the simulates a sky quality meter. This device measures brightness of the sky at the zenith and displays a value in units of magnitude per square arcsecond (mag/arcsec^2).

There are several web sites that publish light pollution maps based upon these types of measurements. They make it easy for the Outreach Astronomer to research and find good locations for an event:

Clear Dark Sky: https://www.cleardarksky.com/maps/lp/large_light_pollution_map.html/
Dark Site Finder: https://www.darksitefinder.com/maps/world.html#4/39.00/-98.00
Go Astronomy:https://www.go-astronomy.com/dark-sky-sites.php
Light Pollution Map: https://www.lightpollutionmap.info/s/najNMssEqphO0XIhlSxw
National Park Service" https://www.nps.gov/subjects/nightskies/

4.4.1   School and Public Park Events

Setting up and operating telescopes is best done on level, hard surfaces. Basketball courts, parking lots, and compacted sports fields generally work well. If the event is to take place on a sports or grass field – make sure that the sprinklers have been turned off. Many outreach events have been interrupted or ended prematurely because an irrigation system turned on at the most importune time.

If you are planning to set up in a parking lot, be sure to address safety concerns. Cordoning off the area before anyone arrives is the first step to ensure safety. Placing cones and caution tape around the area will also ensure that the space is available when needed for setup.

Clearly communicate the time for setup and tear down with your volunteers and staff. Astronomers should be instructed to never drive their vehicle into the area after an event has started. Likewise, the teardown time should be strictly enforced to ensure safety. Keeping vehicles away from the area protects participants and reduces the possibility that headlights will ruin the viewing during the event.

4.5   Event Scheduling

Telescopes can be used on almost any date assuming weather cooperates. Many astronomers prefer nights close to or during a new moon to support viewing of dimmer objects. But planetary viewing usually doesn't suffer greatly even during a full moon. If the goal is to observe the Moon itself, the more it is illuminated (e.g., closer to full phase), the harder it is to view in a telescope. It becomes too bright and uncomfortable to observe. A lunar filter (aka polarization filter) can help but fewer details will be visible because of a lack of shadows and contrast.

Lunar viewing is best done during a waxing crescent, first quarter, third quarter, or waning crescent phase. This is when shadows reveal more details on the lunar surface. Observing along the terminus (the line dividing day and night) can also offer better views of craters, mountains, and rifts.

Appropriate dates and times will be dictated by the calendar. Daylight savings Time (DST) can greatly limit participation especially by younger audiences. Generally, teachers and parents prefer that their kids get to bed early on school nights. Starting a 2-hour event at 9:00 pm would not be a good idea for elementary school students during the school week.

Fig. 4.3 provides suggested start times for telescope events that occur during Daylight Savings Time. Sunset times listed are estimated and will differ by location.

In general, elementary school events should start by 7:00 pm and the duration should be limited to 2 hours or less. Accordingly, elementary school events are typically not held during Daylight Savings Time. The best dates to hold an elementary school event fall outside of DST between the first week of November and the first week of March.

Selecting appropriate dates for an event can also often be limited by object viewability. If a goal is to offer viewing of as many planets as possible – then research will be needed to determine when the planets of interest will be visible during the time of the outreach event. Fortunately, there are many very good tools available to help with this planning.

Fig. 4.3 Suggested event start times during Daylight Savings

DATE	APPROX. SUNSET	VIEWING START TIME
Mar 18	6:56 PM	8:00 PM
Apr 3	7:08 PM	8:15 PM
Apr 18	7:20 PM	8:30 PM
May 3	7:36 PM	8:45 PM
May 18	7:48 PM	8:45 PM
June 3	7:58 PM	9:00 PM
June 18	8:05 PM	9:00 PM
July 3	8:06 PM	9:00 PM
July 18	8:02 PM	9:00 PM
Aug 3	7:50 PM	8:45 PM
Aug 18	7:35 PM	8:30 PM
Sept 3	7:15 PM	8:15 PM
Sept 18	6:54 PM	8:00 PM
Oct 3	6:33 PM	7:30 PM
Oct 18	6:14 PM	7:15 PM

Figure 4.4 lists five on-line tools that are commonly used to plan dates for astronomy events. The table indicates which tools provide useful information. For example NASA's Spot the Station at https://spotthestation.nasa.gov/ and Heavens Above (https://www.heavens- above.com/) web sites both provide viewing dates and times for the International Space Station (ISS) based upon location.

The Heavens Above web site also lists rise and set times for planets by date. This is handy information for when a school asks for a list of possible dates to hold an event. By considering sunset, day light savings, lunar phase, and object visibility, the outreach astronomer can easily find dates to recommend.

In addition to these on-line resources, many planetarium software packages include a function to list best-viewing opportunities for objects. This can also be helpful when researching what is visible on any given date and time. I often utilize the on-line sites in Fig. 4.4 to develop a target list of dates and objects. Then I confirm the list using planetarium software. The software also helps visualize how the evening observing program will unfold.

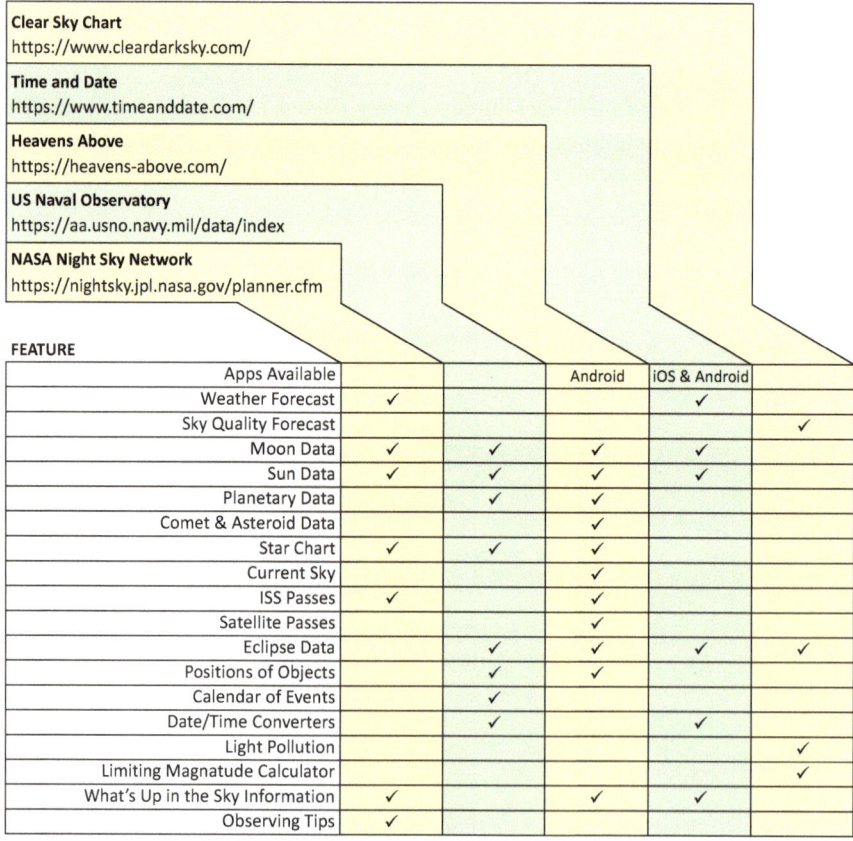

FEATURE	Clear Sky Chart https://www.cleardarksky.com/	Time and Date https://www.timeanddate.com/	Heavens Above https://heavens-above.com/	US Naval Observatory https://aa.usno.navy.mil/data/index	NASA Night Sky Network https://nightsky.jpl.nasa.gov/planner.cfm
Apps Available		iOS & Android	Android		
Weather Forecast		✓			✓
Sky Quality Forecast	✓				
Moon Data		✓	✓	✓	✓
Sun Data		✓	✓	✓	✓
Planetary Data			✓	✓	
Comet & Asteroid Data			✓		
Star Chart			✓	✓	✓
Current Sky			✓		
ISS Passes			✓		✓
Satellite Passes			✓		
Eclipse Data	✓	✓	✓	✓	
Positions of Objects			✓	✓	
Calendar of Events				✓	
Date/Time Converters		✓		✓	
Light Pollution	✓				
Limiting Magnatude Calculator	✓				
What's Up in the Sky Information		✓	✓		✓
Observing Tips					✓

Fig. 4.4 Comparison of on-line astronomy planning sites

Several free apps are available for smart phones that can also be useful for planning purposes. Some favorites include:

App name	Platform	Information provided
The Moon	(iPhone, Android)	Phases of the Moon.
Heavens Above	(Android)	Sky chart, ISS, satellites.
ISS Detector	(iPhone, Android)	ISS passes by date and time.

Sharing this information with the other astronomers at the event is an important organizational step. It helps to suggest targets to observe to avoid all of the telescopes being pointed in the same direction at the same time. Sending an email with the form shown in Fig. 4.5 to the astronomers before the event can help.

Outreach events held under dark skies are great opportunities to share views of dim objects not easily seen from urban and suburban locations. Observing planetary nebulas (such as the *Ring Nebula*, *Dumbbell Nebula*, large nebulas like the *Swan* or *North America)*, or clusters of galaxies like *Stephan's Quintet* are exciting and rare opportunities for guests. They also serve an important educational purpose. If a

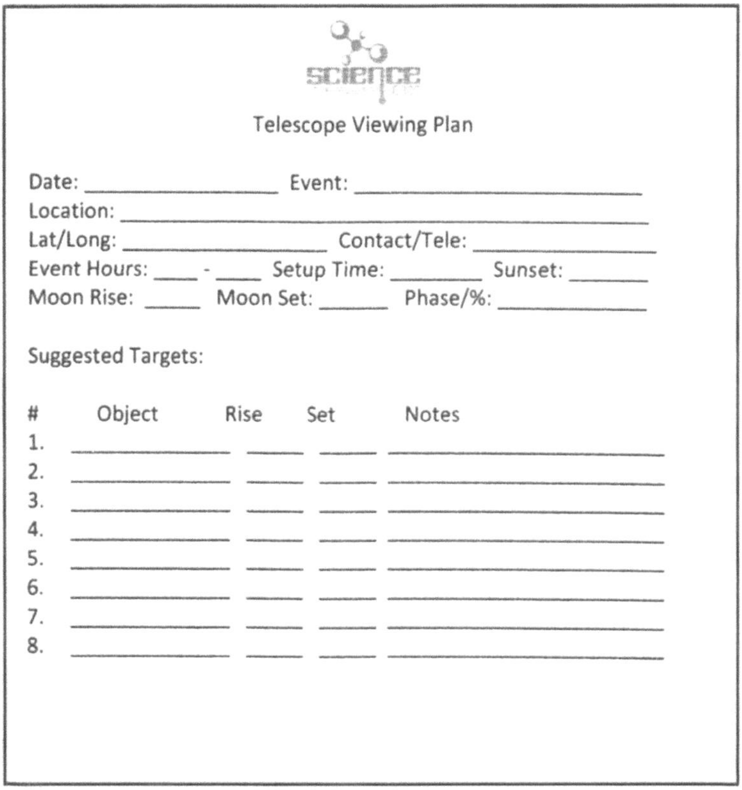
SCIENCE

Telescope Viewing Plan

Date: ______ Event: ______
Location: ______
Lat/Long: ______ Contact/Tele: ______
Event Hours: ____ - ____ Setup Time: ______ Sunset: ______
Moon Rise: ____ Moon Set: ____ Phase/%: ______

Suggested Targets:

#	Object	Rise	Set	Notes
1.				
2.				
3.				
4.				
5.				
6.				
7.				
8.				

Fig. 4.5 Example of a telescope viewing plan form

telescope with an appropriate aperture and focal length is used, these objects can offer breathtaking views.

Creating an observation plan is easy when using planetarium and observational planning software. Fig. 5.19 in Chap. 5 provides a list of free planetarium software useful for this purpose.

The following is a list of popular observational planning software and sites that are also free:

Astroplan – Open Source, runs on python https://pypi.org/project/astroplan/
AstroPlanner – runs on macOS & Windows https://astroplanner.net/
Deep Sky Planner – runs on Windows http://www.knightware.biz/dsp
Telescopius – web only https://telescopius.com/

The above applications allow you to filter by date, magnitude, start and end time. These settings are useful for planning a schedule for an outreach event. Some of these applications also offer telescope control. Theoretically you could automate your entire event observing program.

4.5.1 Planetary Features

With a bit of planning the outreach astronomer can also be prepared for unique views of the moons and planetary features of Jupiter and Saturn. The best-known feature on Jupiter is the Great Red Spot (GRS). It's believed to be a storm that has been raging in the planet's atmosphere for possibly more than 300 years. While it has shrunk over the past few years and no longer appears to be very red it can add interest and educational value to your viewing event. The GRS is only visible approximately every 10 hours because of the rotation of the planet.

The *Sky and Telescope's* magazine website provides a very useful on-line app that calculates when the GRS will be visible:

https://skyandtelescope.org/observing/interactive-sky-watching-tools/

Identifying Jupiter's larger moons, also known as the Galilean moons, adds useful and interesting information to a viewing session. Again, *Sky and Telescope* offers a very helpful on-line app for visualizing the position of the moons on any given day and time. When using this application be sure to select the correct orientation based on the type of telescope used (Refractor, Newtonian, or SCT) otherwise the positions of the moons may not match what you see in the eyepiece.

Lastly, *Sky & Telescope* also provides an on-line app for visualizing Saturn's moons. None of these applications include a print function. But you can print using your browser or using your computer's screen capture function, then pasting the image into another application that does support printing. Of course, you can also write the information on the planning form provided in Fig. 4.5.

4.5.2   *Meteor Showers, Comets, and Eclipses*

Astronomy outreach events are sometimes scheduled to occur during specific cosmological occasions. Lunar and solar eclipses are two events that logically call for
an outreach to be scheduled. Below are two on-line resources from NASA about
upcoming eclipses that detail the paths of totality:

NASA Catalog of Past & Upcoming Lunar Eclipses: https://eclipse.gsfc.nasa.gov/
 lunar.html
NASA Information About Upcoming Solar Eclipses: https://solarsystem.nasa.gov/
 eclipses/future-eclipses/

Newspapers and TV news programs will frequently raise public awareness several days in advance of special events. With some advance planning, your event can
be part of the excitement. Contact local radio and TV stations at least a week before
your event. Send out press releases to local newspapers and internet influencers. It's
free advertising that can get many people to your eclipse watching event.

Knowing the path of totality for an eclipse is important when selecting a location. Just about any venue that can support a lot of people will usually work well –
schools, parks, libraries, museums, parking lots, grassy fields. All you need is an
unobstructed view of the sky and cooperating weather.

Meteor Showers

Meteor showers are favorite times to host nighttime outreach events. Coordinating
a star party during a meteor shower can add excitement to your entire event. Hearing
the "oohs" and "wows" when a bright meteor trail appears overhead is always thrilling. When advertising a meteor shower event be sure to set people's expectations
appropriately. Light polluted urban and suburban skies severely limit what can be
seen. In many instances, events have led to disappointment because expectations
were set too high. Figure 4.6 lists the months of the major meteor showers. Check
with astronomy web sites and magazines and NASA's Night Sky Network website
for best days and times to view the shower.

Fig. 4.6 Meteor showers
by month

Meteor Shower	Month
Quadrantids	December / January
Lyrids	April
Perseids	August
Orionids	October
Leonids	November
Geminids	December

Comets

Occasionally, comets will make bright appearances. Comets often brighten as they get closer to the Sun and on rare occasions, they brighten enough to be seen with just the naked eye. Care must be taken to view comets safely during the day because of their proximity to the Sun. Telescopes and binoculars are not recommended because of the risk to the viewers. Outreach astronomers can help guide guests on how to view these objects safely.

Nighttime viewing of comets requires a telescope, although several comets over the past few decades have been easy to spot with just binoculars. A comet viewing event presents a fun and educational opportunity for both children and adults. If the viewers are lucky, they may see two tails – a trail of debris opposite the comet's direction of motion, and another composed of ions energized by UV radiation emitted by the Sun. The ion trail will point in the direction of the solar wind.

It can also be educational to tell stories about comets. The most famous comet, Halley's, caused worldwide panic in 1910. It was predicted to either crash into Earth or make a very close pass that year. Some scientists at the time predicted that the comet's cyanogen laden tail would engulf the entire planet extinguishing all life.

As a result, many unscrupulous people sold anti-comet pills and gas masks. The only comet related death reported that year sadly was that of famous author and humorist Samuel Langhorne Clemens (aka Mark Twain). He coincidentally was born in 1835 when Halley's comet had made its previous appearance. Clemens had predicted some years prior that his own demise would occur when Haley's comet returned, and he was correct.

Historically comets were thought to be harbingers of bad things to come. Their appearance would cause kings to hide in fear. Wars were fought. Plagues and firestorms in London reportedly followed the appearance of a comet in 1664.

Even in modern times comets have been associated with sad and unfortunate human behavior. In 1997 the Heaven's Gate cult believed that the Hale-Bopp comet appearing that year was hiding a spaceship coming to take their members away. Their belief was so strong that they committed mass suicide.

It may be very easy for us to think in clinical terms about astronomy and celestial events. But the fact is throughout history what happens in the sky often motivates people to do things; some of which would seem to be irrational and counter intuitive. Sometimes their actions were based on ignorance, sometimes based on science. The unknown can be very scary. This is one reason why the role of the Outreach Astronomer is so important.

Refer to national, local, and astronomy news outlets for specific viewing information and forecasts for comets.

Eclipses of the Sun

Partial, annular, hybrid, and total solar eclipses are unique not just as cosmological events but also as social and educational events. Special considerations are required to safely conduct outreaches during these events. Looking directly at the Sun can cause immediate and non-reversable damage to the human eye. To safely view a solar eclipse, you and your guests will need solar eclipse glasses, welder's goggles, or solar filters. Pin hole cameras are another way to safely view an eclipse.

While eclipses may occur somewhere on the Earth almost every year, they rarely occur where people happen to be. In most cases people must travel great distances to get into the path of totality When it does happen it provides a remarkable opportunity to educate people about eclipses and the Sun. And the experience itself is truly stunning and unforgettable.

Telescopes are not required to view an eclipse and first timers should put them aside in favor of simply using eclipse glasses. Viewing the event with proper eye gear is the most rewarding way to observe an eclipse. Those lucky enough to have experienced eclipses before, may want to try their hand at capturing details of the event with cameras. But everyone else should focus on the moment and the visual splendor of the event and its phases as shown in Fig. 4.7.

A total solar eclipse from start to finish can take an hour or more. Totality itself can last from just a few seconds to up to 7 and half minutes. The duration will depend on many factors including where the viewer is in relationship to the center line of the path of totality.

The five phases of a total solar eclipse are described below. The phases are referred to as 1st, 2nd, 3rd and 4th contact. Between 2nd and 3rd contact is the period referred to as totality.

The safety procedures outlined below apply only to total solar eclipses. It is important to understand that it is never safe to view the Sun without specially manufactured filters and glasses examples of which are shown in Fig. 4.8.

1st Contact – The beginning of the eclipse. Before and during this phase the use of safety filters and glasses is mandatory. Viewing the Sun directly can cause permanent eye damage. Outreach Astronomers should warn unprotected and unprepared viewers of the danger involved. Telescopes fitted with solar filters and h-alpha telescopes can also be safely used during 1st contact.

If you are situated on top of a hill, prepare yourself to see the shadow of the Moon racing toward you. It will be traveling at around 1 km per second and will appear very quickly in the moments leading up to 2nd Contact.

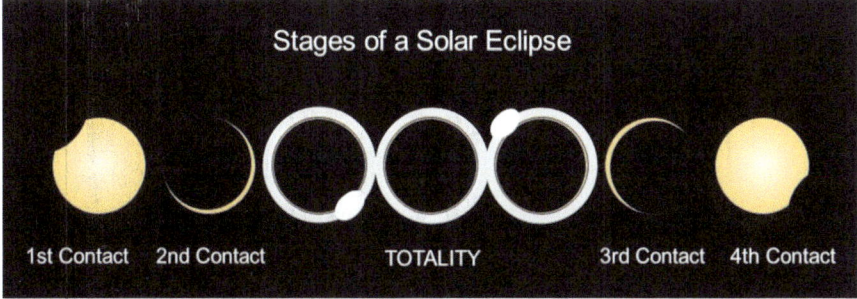

Fig. 4.7 Five stages of a total solar eclipse

2nd Contact –In the moments leading up to totality, the Sun is nearly completely blocked by the Moon. Special effects known as the "Diamond Ring" and "Baily's Beads" will appear. Shadow bands may become visible on the ground and the walls around you. Solar glasses and filters are still needed at this stage. Telescopes will be useless as too much of the light is being blocked by the Moon to get through the telescope's filters. Concentrate on looking at the Sun using your solar eclipse glasses and filters.

Totality – During totality (total solar eclipses only) the Sun's disk is completely blocked by the Moon. It is generally safe to view the Sun at this point without using solar eclipse glasses (again not for annular or hybrid eclipses – only total solar eclipses).

Totality brings near darkness to the environment around you. At this point you are engulfed in the Moon's shadow. The air temperature drops several degrees. The colors of objects around you seem to change. Animals react to the sudden darkness: Cows may return to their barns, crickets will start chirping, chickens and other birds return to their roosts. People around you will react vocally.

Around the edges of the blocked Sun, you will see the corona and streams of solar wind heading out into space. Prominences emanating from the chromosphere become visible. Planets, stars, and comets near the Sun also become visible.

3rd Contact – The moment totality ends your guests should be instructed to put their solar eclipse glasses on again. Outreach astronomers should confirm that their guests are viewing the Sun safely. The "Diamond Ring" and "Bailey's Beads" may appear again. The Moon's shadow is now rushing away from you.

4th Contact – The eclipse ends. It's time to catch your breath and look around. Your guests may display many emotions and reactions. Ask them to describe what they saw. Putting it into words can be hard for some. Doing so will help them to remember the event. It's also not too early to start planning for the next eclipse. In the hours that follow write down what you would do differently next time.

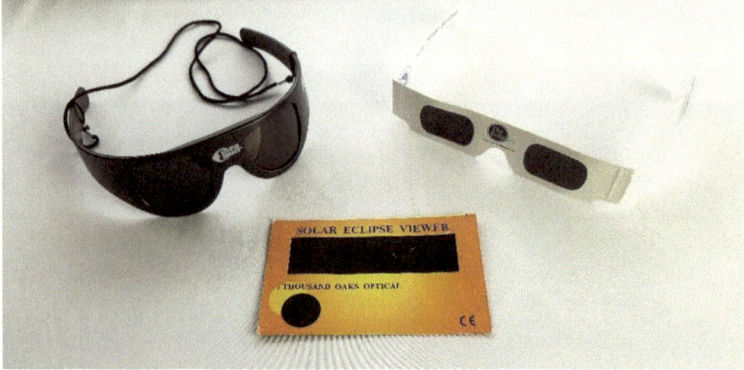

Fig. 4.8 Commercially available solar eclipse glasses and filters

Science During Solar Eclipses

Historically, eclipses have been used by scientists to perform experiments. Perhaps the most famous scientific discovery made during a total solar eclipse was the confirmation of Einstein's Special Theory of Relativity. In 1905, Albert Einstein published three papers in the German science journal *Annalen der Physik*. Up to this point, Einstein had been working as a Swiss patent examiner and was unknown to most scientists.

One of his three papers, the Theory of Special Relativity, and the principle of equivalence he later described, implied that the path of atomic particles and light would be bent in the presence of a very strong gravitational field such as that produced by the Sun. Einstein realized that during a total solar eclipse it would be possible to test his hypothesis. By comparing the actual and apparent positions of stars visually near the Sun, astronomers could determine if his theory was correct.

World War I intervened, though, and this experiment would not be possible to perform until 1919. In that year, two teams of British astronomers, one headed to Brazil the other to an island off the coast of Portugal, set off to test Einstein's theory. The world would learn of the results of these experiments on November 6, 1919, when they were presented at a joint meeting of the *Royal Society* and the *Royal Astronomical Society*.

Einstein's theory was proven correct. Light is bent by strong gravitational fields. Practically overnight Albert Einstein, a once obscure patent clerk, became the world's most famous scientist and our understanding of the Universe profoundly changed. Since then, space-based telescopes have produced amazing images of so-called Einstein rings exhibiting the effects of gravitational lensing. This phenomena, illustrated in Fig. 4.9, is now being used to map dark matter in our Universe.

4.5.3   International Astronomy Observances

2009 was the *International Year of Astronomy* (IYA). It was a year- long celebration coinciding with the 400th anniversary of the beginning of modern astronomy. In 1609, Italian mathematician Galileo Galilei turned his newly acquired, newly invented, telescope toward the night sky. Accordingly, Galileo is considered to be father of modern astronomy, physics, and the scientific method.

The IYA was organized and supported by many professional and amateur astronomy groups, organizations, and governmental agencies including the International Astronomical Union (IAU), NASA, the European Space Agency (ESA), and many others. The celebration was enthusiastically embraced by thousands of scientists and amateur astronomers worldwide. Activities hosted during the year ranged from having children build simple refractor telescopes, to public presentations and Q & A sessions at libraries, museums, observatories, and schools. For many this was their first opportunity to interact with researchers many of whom were working at

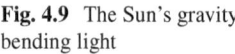

Fig. 4.9 The Sun's gravity bending light

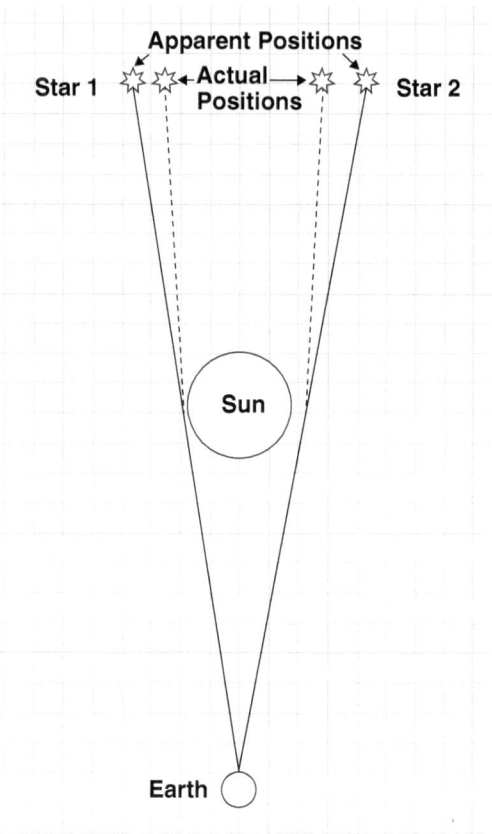

the world's most advanced observatories. The public got to see science in action. The IYA was both breathtaking and groundbreaking in its scope and impact.

The success of the IYA spawned many efforts that continue to this day. Some current programs have the objective of inspiring kids to study STEM subjects; others are designed to encourage diversity; many seek to build public support for research and protecting the night sky.

Organizing your own outreach events around these on-going programs is a good way to boost interest and participation. These international efforts can be effect multipliers for your own event. Some of the best-known on-going programs are:

- **International Astronomy Day** (occurs twice a year during the first quarter moon in the spring and fall).
- **Global Astronomy Month** (promoted by Astronomers without Borders and is held every April. See https://www.astronomerswithoutborders.org/)
- **United Nations International Asteroid Day** (anniversary of the most harmful, recent asteroid impact – the Siberian Tunguska event, June 30th. See https://www.un.org/en/observances/asteroid-day).

- **United Nations International Moon Day** (July 20. See https://un.org/en/observances/moon-day).
- **NASA's International Observe the Moon Night** (every October. See https://moon.nasa.gov/observe-the-moon-night/).

4.5.4 Citizen Science

There are many opportunities today for non-scientists to participate in real scientific research. NASA hosts a site specifically listing many of these opportunities at https://science.nasa.gov/citizenscience. Projects include studies of the Earth's clouds and atmosphere, asteroids, minor planets, exoplanets, stellar dust disks, dark matter, and dark energy. They are all great learning opportunities for anyone with the interest and time to participate.

Another type of research that can be combined with an astronomy outreach is the observation of occultations. The term occultation refers to when one object passes in front of another – either partially or completely blocking the light from the more distant object. The simultaneous and accurate timing of occultations from different locations on the Earth has enhanced our knowledge about many objects.

Sharing a live view of an occultation is a great way to demonstrate science in action. You can explain how teams of astronomers, located in many different places, are timing the event. The data collected will be analyzed to reveal details about the observed object. For example, it can reveal the shape of an asteroid and its rotation rate.

The International Occultation Timing Association (IOTA) was formed in 1983 specifically to assist amateur and professional astronomers with the task of collecting occultation data. Through their efforts, astronomers have been able to fine tune their understanding of the shape and orbit of asteroids, the topography of the Moon's polar regions, and even identify binary stars. IOTA's web site (https://occultations.org/) provides information about the technique, the tools needed, and lists the opportunities to participate.

4.5.5 Heritage Awareness Celebrations

There are a large number of awareness days, weeks, and months that are recognized around the world. The United Nations lists hundreds of specific days and weeks that celebrate various languages, cultures, and efforts at https://www.un.org/en/observances/list-days-weeks.

In the United States over a dozen month long celebrations are recognized for various groups. Coordinating themed outreach events with these celebrations is a great way to multiply the impact of your event. For example, giving a talk about the contributions made by African American scientists during Black History month will generate interest and participation. Frequently national and local media run stories

Fig. 4.10 List of heritage and other celebration months

CELEBRATION	MONTH
Black History	February
Women's History	March
Irish American Heritage	March
Greek American Heritage	March
Arab American Heritage	April
Asian & Pacific Islander Heritage	May
Jewish American Heritage	May
LGBTQ+ Pride	June
Caribbean American Heritage	June
Immigrant Heritage	June
Disability Pride	July
French American	July
Hispanic Heritage	September
German American Heritage	October
Filipino American History	October
Italian American Heritage	October
Native American Heritage	November

about these celebrations raising people's awareness and interest in the topic. The purpose of heritage celebration months is to garner appreciation and respect for the individuals, cultures, and groups being celebrated.

You can find many ways to tie these celebrations into an astronomy event. Consider the linguistic source of star names. Hundreds of stars have Arabic names (Altair, Betelgeuse, Kochab, Merak, Rigel). Arab astronomers during the 9th - 13th centuries made many important and lasting contributions to astronomy. Likewise, Chinese and Japanese astronomers documented many events that we still refer to today. And of course, there are many constellations named after mythological Greek figures. Heritage celebration months are a natural time to educate the public and at the same time celebrate contributions by various cultures.

As a NASA Solar System Ambassador, I have given several talks timed to coincide with heritage month celebrations. In each case, I found that the free publicity surrounding the month-long celebration added to the event and grew the audience. Figure 4.10 lists many of the better known celebrations.

4.6 Staffing an Event

Unless you are planning the simplest of events, chances are you will need people to help. How much help needed will depend on several factors including the type and size of the expected audience and the variety of activities planned for the event.

Science Heads Inc. regularly hosts astronomy outreach events that draw between 200 and 800 people. The number of participants often depends on the size of the school or the community where it is being held. A typical event for Science Heads

Inc. includes setting up a mobile observatory along with a dozen or more hands-on activity tables. Volunteers are needed for setup, tear down, and running the activities. We often operate alongside astronomers from local astronomy clubs. Events are frequently part of a larger school event such as a Family Science Night.

Handling large crowds and being effective informal educators takes a lot of effort and many volunteers. Consider the efficiency of a simple star party event. One lone telescope operator can often accommodate up to 100 people over a two-hour period. With that number, each guest would be allowed a little over 60 seconds at the eyepiece. Add in the time to slew to different objects, make various adjustments, eyepiece changes, focusing, etc. and the observation time for each person may drop significantly. Even a full minute doesn't seem adequate for the astronomer to explain what's being observed and allow the viewer to take in all the details with their eye.

Supporting 100 guests per telescope produces long lines and long wait times. A typical wait at the above-described telescope could probably be 30 minutes or longer. Some guests may give up before they reach the telescope. Waiting this long for a short observation opportunity is sure to invite dissatisfaction. To keep wait times down to 15 minutes or less more telescopes and astronomers are needed. A better ratio at an event maybe two telescopes for each 50 or 100 participants.

To keep guests interested, Science Heads Inc. provides educational opportunities for guests while they are waiting on the line. This requires assigning one or more volunteers to engage with the guests, answer questions, explain what is being observed, even handing out "prizes". We provide the volunteers scripts and trivia questions appropriate for the age and grade levels.

The above ratios work well for standalone telescopes. Using fixed and mobile observatories as described in Chap. 5 requires a very different calculation. Observatories can offer more activities than just viewing through a telescope. In Science Heads' mobile observatories there are interactive workstations, videos, in-person discussions, and Q & A sessions offered. The amount of time each guest will spend in the observatory increases dramatically compared to a standalone telescope. Guests may spend between 15 and 30 minutes or more inside. This participation time must be factored into the planning for the event.

Staffing for an event may also require other roles be filled besides telescope and observatory operators. If hands-on activities are provided, volunteers are needed to set up tables, tents, and to run the activities. Teardown after the event requires staffing. Volunteers may be needed in the parking lot to direct traffic, answer questions at an information table, and monitor the event for safety.

To determine the appropriate staff for an event first list all the tasks that need to be done before, during, and after the event. Using the form in Fig. 4.1 can help. Then considering the size of the audience expected, estimate how many volunteers are needed for each task. This will be your recruiting goal for the event. Lastly assign the tasks to your staff and volunteers and make sure that everyone knows their role and responsibilities.

4.7 Marketing and Advertising

There is no doubt there will always be a role for low budget, even spontaneous astronomy outreach events. Setting up a telescope on a street corner or at a park and sharing views of Saturn and Jupiter is not just fun but educational. The *Sidewalk Astronomers*, started by John Dobson in 1968, promotes and engages in this type of outreach activity. Dobson believed in setting up telescopes where people gathered and showing them that expensive equipment is not necessary. Dobson also regularly showed amateur astronomers how to polish their own mirrors and designed what we now call the Dobsonian telescope.

There is also a place for larger and more involved outreach events as discussed in this book. All these types of events can likely benefit with a bit of marketing and advertising. After all, if you are going through the effort, why not get as many people there as possible?

Marketing is the process of turning objectives into reality. A good marketing plan converts the who, what, when, where, and why into event details and communications that will achieve the stated objectives of the event. You can think of an outreach event as a product. Marketing professionals plan what they call the four P's: Product, Production, Promotion, and Price.

In terms of an outreach event the Product is the event itself. What will be happening at the event (a star party, hands-on activities, presentations). Production is the when and where (location, date, and times). Promotion is the communication tools used such as flyers and advertisements. And Price – well that is the cost, if any, of participating in the event. Manufacturers spend a lot of time calculating the cost of a product and the price at which to sell it. Outreach astronomers generally don't need to worry much about price since most events are free to the public.

An advertising plan on the other hand identifies the specific media outlets that will be used for event advertisements, when the ads will be run, and for how long. Communications that are part of an advertising plan can take forms other than advertisements. Flyers, hand-outs, web pages, emails, text messages, and more may be part of the advertising mix.

Both marketing and advertising plans should explain the rationale for each component of the plan. The objective may be to support STEM education at local schools. Accordingly, a marketing plan may be defined to get local families to come to the event. The plan will describe the messages that will interest and motivate families to participate. The advertising plan will identify the specific media outlets that will reach this intended audience based upon viewer demographics and distribution.

A family-oriented astronomy event plan may identify the following motivators that can be promoted:

• The event is family friendly.
• Elementary school children will learn about STEM subjects.
• The event is free.
• It's at your student's school.

- There will be many children's activities.
- The event occurs on a day and time that often families are looking for something to do (ex. Saturday afternoon).

Conversely, an event that is targeting primarily adults will likely identify different factors:

- A renown or authoritative presenter is giving the talk.
- It's being given at your local library, museum, or city hall.
- The event occurs on a day and time appropriate for adults.
- Topics covered are appropriate for an adult audience.
- A Q & A session will be held after the presentation.
- Materials and links for continuing education will be provided.

To reach the first audience the advertising plan may include:

- Distributing flyers at schools and libraries.
- Posting on social media sites that target parents.
- Running ads in the local newspaper.
- Adding events to on-line "what to do this weekend" calendars.

To reach the second audience the plan may include:

- Putting up notices at libraries, coffee shops, stores, city hall.
- Posting on astronomy and STEM related discussion groups.
- Running ads on radio and TV.
- Posting flyers on college campuses.

The respective advertising plans differ because of the differences between the two intended audiences. Adults are more likely to go to a coffee shop, city hall, and listen to the radio. Students on the other hand are likely to learn about events from their teachers.

People have come to expect a certain level of professionalism even for community events. The quality of marketing materials generally demonstrates the quality of the program being offered. While outreach astronomers may not feel the need to sell their event – in today's commercialized world professional looking materials are expected and absolutely needed. We are all bombarded with advertising from many sources. Everyday emails pour into our inbox. We may receive and ignore hundreds of communications every week. Getting above the din and noise of all that advertising is needed for any type of event.

A typical event marketing and advertising plan will include:

- A budget for printing, distribution, advertising.
- Copy (aka text) that describes the event for use in flyers, posters, signs, advertisements, press releases, sponsor appeals, etc.
- Schedules for running advertisements, distributing flyers, contacting media.
- Rationale for ad placement and material distribution.
- Description of the mechanism to sell tickets, the price for the event, how attendance will be managed.

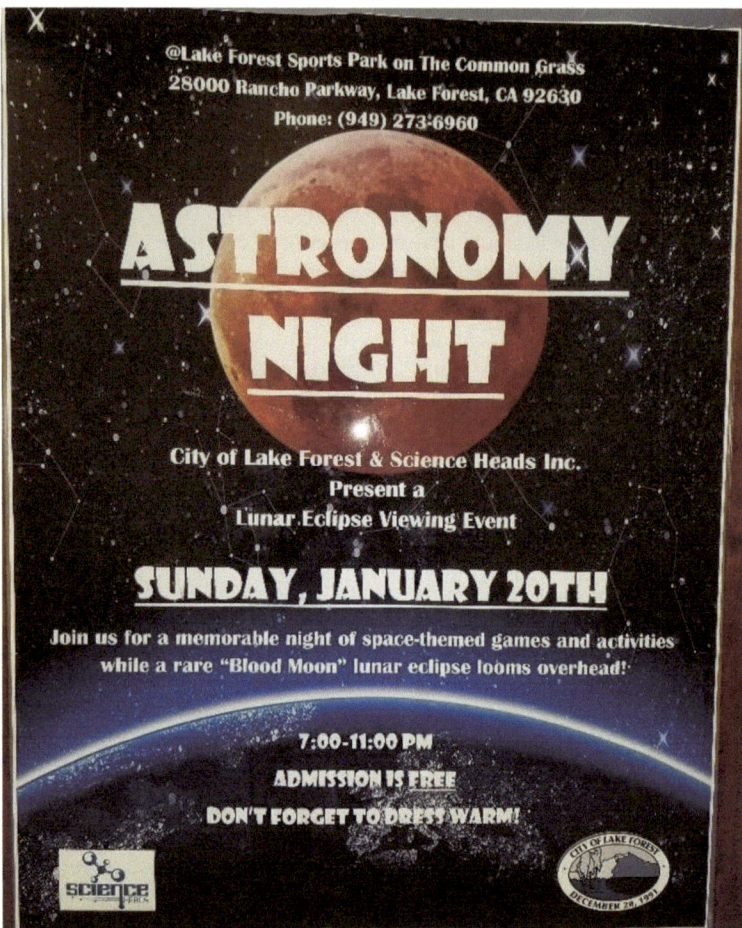

Fig. 4.11 Outreach event poster

- Identify who will do each marketing task and when.
- The metrics and tools that will be used to measure progress and success.

Figure 4.11 shows an example of a city park event poster.

Fortunately producing clean and professional-looking posters and flyers is easy with applications readily available on personal computers. Try to keep your designs clean and easy to read. Limit the information to the basics. Unnecessary information and a very busy design may cause the poster to be ignored. Generally, people look at three places on a poster and then quit reading. Try to keep your design simple and fast to read. Add a QR code that links to a web page so readers can get more information. It encourages people to take a picture using their smart phone, so they don't forget.

There are also many affordable on-line social media sites that accept advertising. Some of the sites listed below support targeting by city, zip code, region, and viewer demographic. Demographic data includes age, presence of children in the household, household income, and other useful parameters.

Facebook https://www.facebook.com/business/ads
Google https://ads.google.com/
Nextdoor https://business.nextdoor.com/
The Patch https://info.patch.com/classifieds/
X (Twitter) https://business.twitter.com/en/campaign/welcome-to-twitter-ads.html/
Yelp https://business.yelp.com/products/yelp-ads/

And don't forget to search for free advertising opportunities that are available in your area. Many TV and radio stations offer "Things to Do Calendars" and free public service ads (PSAs) for non-profits. Local newspaper websites, community cable channels, and community websites offer free postings for local events. Some radio and TV stations and on-line influencers might even be interested in interviewing you on the air. Newspapers, TV and radio stations, bloggers, and influencers are often looking for interesting stories to cover.

Commercial advertisers know that creating customers requires repeated exposure to their advertising. It is generally understood that between 7 and 50 impressions of an advertisement is needed before a viewer becomes a customer. This is why you see so many ads repeated on television.

Many businesses want to support local non-profits. They may be willing to post posters and flyers at their place of business. Coffee shops, donut shops, restaurants, and stores may offer free space for this purpose. Always get permission from the manager before hanging a flyer or poster and be prepared with all the materials needed: thumbtacks, tape, display boxes, business cards.

Marketing does not end once an event starts. A good marketing plan defines "touch points" with the audience before, during, and after an event. The goal of a marketing plan is to maximize impact and outcome. This involves building relationships with participants. Businesses recognize that their most likely future customers are their current customers. Likewise, the most likely person to come to your event is someone who has participated in a past event. Keep these people informed about what you are doing and when your events are being held. They may also provide some of your best advertising through word-of-mouth.

Keeping in touch with your audience is also important to maintain a connection with your community. Setting up a Facebook page, an X (Twitter) account, Instagram, TikTok, Discord, etc. are good ways to keep them involved. An organization newsletter (print or on-line) is also useful to engage your followers. If you send emails, it's best to follow industry best practices for opt-in and opt-out. Let your followers check the box to opt-in to the mailings. It ensures that you will not get identified as a spammer and keeps your email list current.

There are several cloud-based Contact Resource Management (CRM) and emailing sites that offer free or low-cost services to non-profits. CRM software is designed to help you maintain your lists and automate your communications. They are capable of handling very large lists. They also automate the opt-in and opt-out function freeing you to focus on the message and not the mechanism.

SERVICE	LINK
ACT!	https://www.act.com/
Brevo	https://landing.brevo.com/
Constant Contact	https://www.constantcontact.com/
Mailer Lite	https://www.mailerlite.com/
Mail Chimp	https://mailchimp.com/
Send Grid	https://sendgrid.com/
Vertical Response	https://verticalresponse.com/

Fig. 4.12 Selected CRM and email marketing services

CRM software is a great way to maintain information about your audience and your marketing efforts. It allows you to create campaigns to advertise your events and send out emails and newsletters on a regular basis.

The companies listed in Fig. 4.12 offer CRM and emailing functions. Many of these services can be integrated into your website, for example, allowing your visitors to sign up for a newsletter themselves.

4.8 Weather and Cloud Forecasts

It is important to monitor weather conditions leading up to the start of an event. The safety of all participants should be of primary concern. Keeping the host and venue appraised of the weather will help them tailor expectations of the participants and adjust if needed. If it is forecasted to be cloudy, accommodations may be possible by adding other activities as substitutes for telescope viewing. In some cases, events may need to be cancelled because of the potential for dangerous weather conditions such as lightning storms. The sooner the host is appraised of the forecast, the sooner a decision can be made, and the participants notified.

There is no shortage of sources for weather forecasts. Fortunately, the science of meteorology has progressed to the point that short term weather forecasting has become very accurate. Forecasts are usually applicable to a 50- mile radius of the identified location. Figure 4.13 lists several reliable sources for short term and extended U.S. forecasts.

Knowing if it is going to be cloudy or clear is harder to predict. Cloud coverage and seeing conditions involve many factors related to the microclimate of the immediate area. Cloud coverage forecasts are much less reliable at present than general weather forecasts. Figures 4.14 and 4.15 lists sites and free smart phone apps that provide forecasts for cloud coverage.

WEBSITE LINK	Description
https://www.weather.gov/	National Weather Service
https://weather.com/	The Weather Channel
https://www.wunderground.com/	Weather Underground

Fig. 4.13 Weather forecast links

WEBSITE LINK	DESCRIPTION
https://www.cleardarksky.com/csk/	Clear Sky Chart
https://graphical.weather.gov/sectors/conus.php?element=Sky	NOAA/NWS graphs showing cloud coverage

Fig. 4.14 Cloud coverage forecast links

SMART PHONE APPS	SOURCE	PLATFORM
Astropheric	Apps Store, Google Play	Android, iPhone
Clear Outside	Apps Store, Google Play	Android, iPhone
Meteoblue	Apps Store, Google Play	Android, iPhone

Fig. 4.15 Free smart phone apps for forecasting weather and cloud coverage

4.9   Planning the Layout for an Event

Planning an event can include mapping out where everything should be setup. A simple drawing should be created identifying where the event activities will be located, how power will be distributed, where guests enter and exit the event, the location of restrooms, etc. Call outs should include where tents are needed, lights set up, power required, and audio components positioned. A drawing of this type can help during setup but also may identify potential problems long before the event starts. For example, an activity table near a restroom could interfere with the line for the restroom. The location of power outlets and generators will determine the length of power cords needed and where distribution panels should be located.

To prepare for an event, a pre-event site visit may be required to determine the exact location of utilities and facilities. Knowing these details will help you avoid unexpected obstacles far in advance of the start of your event.

Figure 4.16 shows an example of a layout that was done for a recent Science Heads Inc. event. Note that the locations of all the activity tables are identified. The tables that require power and lighting are also noted on the drawing.

Safety zones on the drawing are delineated with a dotted line. These areas were cordoned off with caution tape. Even the path that led guests to the event was marked (lower left). Immediately adjacent to this path is an information and check-in table. Every member of the setup team was given a copy of this drawing before setup started.

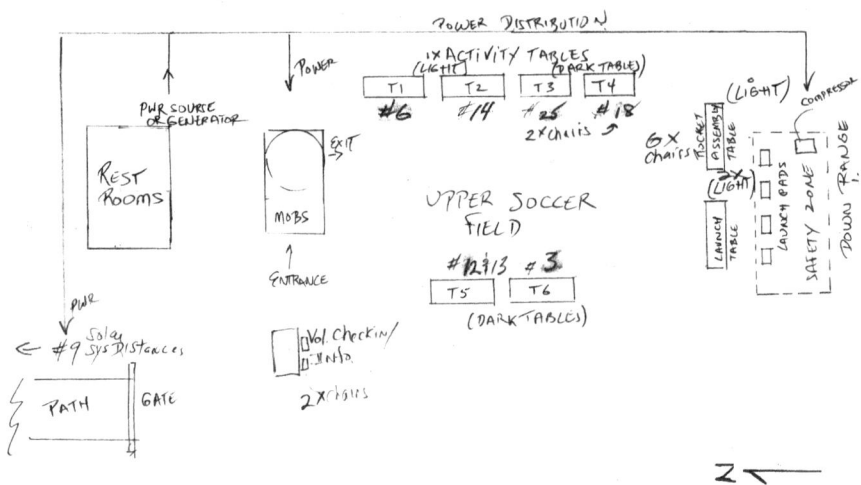

Fig. 4.16 An event layout drawing

4.10   Running an Event

The better an event is planned, the easier it is to run. Unexpected things can always occur, but good planning makes dealing with these contingencies easier. Being prepared, flexible, and ready to deal with the unexpected is key to running a successful event.

The following suggestions may help you run your event smoothly:

1. Before the day of the event think through what is needed for a successful event. Prepare a checklist for the items that you need to bring. Identify the tasks that are yet to be done. Check each off as you load the items into your vehicle or complete the task.

2. Leave yourself extra time before the event starts to deal with last-minute glitches. Arriving on-site 1–2 hours before setup time is usually sufficient. It gives you time to check that the venue is ready, gates and bathrooms are unlocked, and access to the site is possible. If anything is amiss, you have plenty of time to contact the venue or host. Figure 4.17 provides a suggested operation checklist. While on-site review the list and check off completed items.

3. When your event starts, stick with the plan as much as possible. This is particularly import when many people are involved. Communicating last- minute changes to your team may not be possible during the event. Assume that everyone is doing what they were tasked to do. Focus on your own tasks and then deal with any glitches that arise. Only involve the staff and volunteers that are affected by the changes that are needed.

SCHOOL EVENT OPERATION CHECKLIST

Before the Event:

- ☐ Confirm that gates are unlocked. Contact host or grounds staff if needed.
- ☐ Confirm that restrooms are unlocked.
- ☐ Determine the best location to set up telescopes, tables, tents.
- ☐ Identify where transport vehicles should be parked.
- ☐ Cordon off transport parking and setup areas.

During the Event:

- ☐ Direct astronomers to transport parking area. Line up cars side-by-side.
- ☐ Identify where astronomers should setup telescopes.
- ☐ Distribute the suggested Telescope Viewing Plan and discuss options.
- ☐ Explain event tear down plan and safety rules. **Emphasize that transport vehicles cannot be moved until after the event ends.**
- ☐ Point out where rest rooms, snacks, water, and other amenities are located,

After the Event Ends:

- ☐ Monitor tear down and removal of transport vehicles.
- ☐ Check in with the host and discuss level of satisfaction, issues, etc.
- ☐ After the last volunteer leaves, police area to ensure nothing was left behind.
- ☐ Send host or venue follow up survey.
- ☐ Send thank you email to volunteers.

Notes:

_____

_____

_____

_____

Fig. 4.17 Checklist for school event

Outreach Astronomer

Mr. Jim Benet – retired aerospace engineer and former Outreach Coordinator for the Orange County Astronomers.

Mr. Jim Benet led the outreach program of one of the largest amateur astronomy organizations in the United States. From 1998 to 2016 he ran a program that averaged 50 events per year at more than 160 local schools, libraries, and parks. During that period, Jim handled almost all the scheduling, coordination, communications, and personally participated in a majority of the approximately 900 events his organization hosted.

Before retiring, Jim was recognized for his incredible effort not just by the Orange County Astronomers, but also by two other famed astronomy organizations. In 2012 Jim was awarded the highest honor of the Western Amateur Astronomers association – the G. Bruce Blair Award - referred to as the "Noble Prize" of amateur astronomy. In presenting the Blair award, the WAA recognized Jim for running a program that reached over 100,000 people.

In 2013 Jim was presented the Clifford W. Holmes Award by the Riverside Telescope Makers Conference (RTMC). The Holmes award was given to individuals who make a "… major contribution to popularizing astronomy."

Jim is also the person who piqued my interest in Outreach Astronomy. It was one of his outreach events that my son and I attended as I described in the preface of this book. Jim showed me how astronomy outreach is done and contributed ideas for this book.

The Orange County Astronomers is a non-profit organization founded to spread knowledge about astronomy. It currently has over 800 members.

The Western Amateur Astronomers association is a non-profit formed by professional and amateur astronomers in the western United States. Its mission is to advance the public's understanding of astronomy.

The Riverside Telescope Makers Conference was an annual event that for over 50 years brought astronomers, telescope makers and vendors from around the country together. RTMC hosted its last event and ended operations in 2019.

Pictures reprinted with permission of Jim Benet.

Chapter 5
Equipment, Tools, and Techniques

The choice of telescopes and tools that an Outreach Astronomer can use at an event is as varied as the many types of astronomy events. Each event may require a different set of tools based upon the objective and the audience expected. Listed in this chapter are common astronomy tools that are useful for outreach events along with suggested techniques for using them.

5.1 Telescopes and Accessories

There are many manufacturers and models of telescopes currently on the market. But far fewer have the attributes which make them useful for serious observing. You have no doubt heard of the suggestions frequently given to first- time telescope buyers. They are often told to select a telescope that is stable and of good quality. Stability refers to the mount and tripod. A telescope that jiggles or bounces when pointed at a target is nearly useless. Many so called "department store" telescopes fall into this category. Quality refers to the quality of the optics. The importance of a well-polished mirror and quality made lenses becomes apparent very quickly when using a telescope. False colors (chromatic aberration), out of focus objects around the periphery (coma), hard to use mounts and focusers, are all potential issues with poorly made telescopes. These factors are also important for astronomy outreach.

There are other considerations that are important to amateur astronomers but may not be as important for outreach events. A large aperture is important to observe dim objects such as distant nebulas and galaxies. In light polluted skies where many outreaches are held this is usually not a concern. The target list for most of our outreaches primarily includes planets, the Moon, and the brighter Messier objects. Large aperture telescopes are great under a dark sky but not so important in the suburbs.

Tracking mounts and "goto" telescopes can be real time savers during outreach events. While a Dobsonian telescope sets up very quickly and is easy to use, the lack of a tracking mechanism means the operator must adjust it frequently.

Equatorial mounts are singularly important for imaging, since the mount mitigates field rotation, but offers no advantages for visual use. A downside of an equatorial mount is the extra parts and weights required. Set up time can also be a bit longer compared to an Alt-Az mount since polar alignment and weight balancing is required.

Refactor, reflector, and catadioptric optical designs (Schmidt Cassegrain, Maksutov-Cassegrain, Schmidt-Newtonian, Schmidt) are all useful for outreach events. The catadioptric designs offer a compact and versatile folded optical tube that can be used on a variety of mounts, piers, and tripods. They come in small and medium apertures making the telescopes generally lite and easy to transport.

Similarly, Newtonians are generally lite weight and come in a variety of apertures. The placement of the focuser on a Newtonian often mandates that a ladder or step stool be employed, particularly for children.

Refractors can offer stunningly sharp views of objects, but larger aperture refractors will be heavier than Newtonians of equal aperture. The bottom line is any stable and relatively good quality telescope can be of use at an outreach event.

The best telescope for an Outreach Astronomer to use is the one that they know the best. Outreach events are what I call "performance science." Outreach Astronomers use sophisticated and expensive scientific equipment "on-stage" to entertain and educate an audience. Like an actor who must memorize their lines for a play, Outreach Astronomers need to be well rehearsed in the use of their equipment. You will be showing an audience amazing views while operating in near total darkness. It is no fun being on-stage fiddling with controls, or looking for the right eyepiece, or trying to remember the steps to slooh your computer-controlled telescope to a new object. You should be able to do all of it practically in your sleep before you attempt your first outreach event.

5.1.1 The Issue of Looking Through a Telescope

It's important to remember that not everyone knows how to look through a telescope. Young children must be taught, and some older adults also have difficulty. It involves a human ability called *proprioception*. Also known as kinesthesia, *proprioception* is a human being's ability to sense the location of the body's extremities, their movement, and actions. As we mature the human mind develops a mental map of its body parts. Under normal circumstances the mind subconsciously knows where every extremity is located.

It's easy to demonstrate this ability. First look around and locate an object that you can easily reach and hold if you wanted – a pen or a coffee mug for example. Now close your eyes and place your right forefinger on your nose. It's easy to do even with your eyes closed. Your mind knows where both your hand and your nose are located even though you moved your hand. Now, with your eyes still closed, use the same hand to reach for the object that you identified earlier.

It's not easy is it? You may have to move your hand around a bit to feel where the object is located. Your mind knows where your forefinger is but the only positional information it has about the object comes from when you saw it earlier. The object you are reaching for is not part of your mind map.

A guest who has never used a telescope before experiences something like you feeling around for that object. But instead of using their hand to find it they are trying to use their head and face. Positioning one eye directly above an unknown location while in the dark is also not that easy!

A natural reflex in this situation is to reach out with your hand. This allows the mind to correlate the location of the hand with the eyepiece. If you hold the eyepiece in your hand, then your mind has two coordinates to use. It then becomes much easier to bring the eye close to where your hand is located. Children often immediately reach for an eyepiece for this reason. But of course, grabbing an eyepiece is not a good thing to do with a telescope. It's likely to move the telescope off target.

One solution is to give your guest something else to hold. Introducing a step stool between the guest and the telescope works well for this purpose. Ask your guest to hold the top bar of the step stool and then place their eye above the eyepiece. Let them feel the rubber guard of the eyepiece on their face and use that to position their eye correctly. I use a simple elevator speech to help:

> Step on up. Two hands here (pointing to the bar), and one eyeball there (pointing to the eyepiece). Let the rubber guard guide you.

The orientation of the step stool and the telescope is illustrated in Fig. 5.1. The guest keeps two hands on the step stool, offering them support as they lean forward. They lower their face towards the eyepiece until they feel the rubber eye guard. Illuminating the telescope with a red flashlight can also help them see where they are headed.

Once they position their open eye near the eyepiece some children and adults may still find it hard to look through the telescope. It's a lot like looking through a drinking straw. If the straw is not aligned just right, you see the walls of the straw and nothing else. The straw needs to be pointed correctly to see light at the other end. But unlike a straw, a telescope cannot be so simply repositioned when it is pointed at an object. Therefore, it becomes a matter of repositioning the guest's body.

It helps here to consider the design of optical lenses. Lenses have an Optical Axis – an imaginary line running through the center of the lens. Additionally different eyepieces exhibit different values of characteristics known as the Exit Pupil and Eye Relief. As shown in Fig. 5.2, the Exit Pupil is the diameter of the focused image that exits the viewer's side of the eyepiece. Eye Relief is the distance from the glass surface that the focused image appears.

Not only must your untrained guest place one eye above the eyepiece, but they must also align their eye perfectly to the Optical Axis. At the same time, they must maintain a distance from the eyepiece approximately equal to the Eye Relief. Without doing so they end up see nothing through your telescope.

Fig. 5.1 Using a step stool
with a telescope

With experience the Outreach Astronomer can get good at recognizing proper alignment of a guest's body. Imagining a right angle drawn on the side of the guest's head may help. Imagine one vector passing through the top of the guest's head' the other vector (90 degrees from the first) emanating from the guest's eye. The head vector should be parallel with the telescope. The eye vector should be at a 90 degrees angle from the vector of the telescope.

It can often help to advise your guest to either bend their neck or their knees to achieve the correct alignment as shown in Fig. 5.3. Bending the neck helps align the vector that should be 90 degrees from the telescope. Bending the knees can help align the vector that should be parallel with the telescope.

There are also other factors that can affect how well people can see through a telescope. A person wearing glasses may not be able to position their eye close enough to the eyepiece to see a focused image. To accommodate people who wear glasses it is generally best to use eyepieces that have eye relief value of 20 mm or greater. This allows most glass wearers to keep their glasses on and still see a focused image. Alternatively, guests can remove their glasses, but they will likely need to refocus the telescope. This will increase the wait time on your line as your next guest will likely need to readjust the focus.

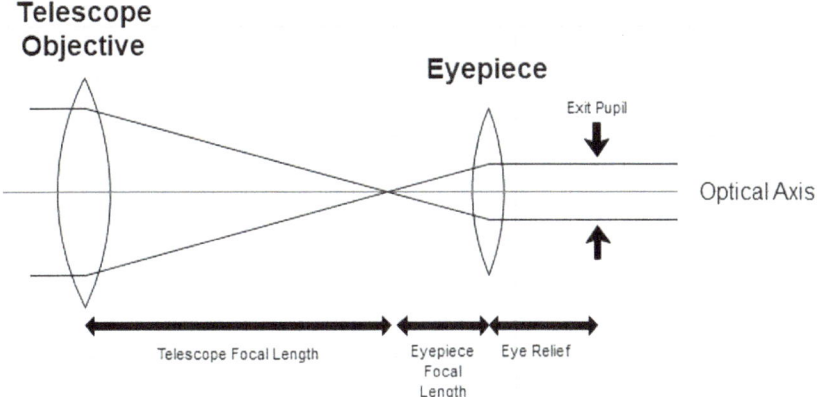

Fig. 5.2 Characteristics of an eyepiece

Fig. 5.3 Bending knees to achieve proper alignment

5.1.2 Dark Adaptation and Averted Vision

Dark adaptation is the natural process of the human eye adjusting to see better in the dark. After some time, the iris of the eye opens wider, letting in more light. Chemical changes also take place in the retina activating the most sensitive components – the rod cells.

As people age, their ability to dark-adapt diminishes a bit. In older people the iris of the eye will tend to open less. All your guests should be encouraged to allow time for their eyes to adapt to the darkness as this will allow them to see more detail in the telescope. The dark adaptation process can take up to 30 minutes. During this time your audience should be instructed to avoid looking at lights other than red lights. Looking at just one streetlamp can reset the whole process.

Another useful technique to teach your guests is how to use averted vision. The cells most sensitive in low light conditions are the rod cells. They are located around the periphery of the retina. In the middle of the retina are the cone cells which detect colors but are not as sensitive to light. Most objects we look at through telescopes don't exhibit color so using rod cells is preferred.

Averted vision is the technique of looking to one side of an object or the other, so the light falls onto the rod cells. We naturally want to look straight at an object so this can take a bit of practice. But by averting one's vision it is possible to see details of an object that weren't apparent otherwise.

5.1.3 Choosing an Eyepiece

The magnification an eyepiece produces is calculated by dividing the focal length of the telescope (in mm) by the focal length of the eyepiece (in mm). Figure 5.4 lists the magnification of common eyepieces. Telescopes with different focal lengths are listed in each column.

When I purchase either a new eyepiece or a new telescope I create and print a similar table, laminate it, and attach it to a lanyard I wear around my neck during outreach events. I find it helps not having to do math while I have so many other

Telescope	f/4 6-inch Newtonian FL = 600 mm	f/6 8-inch Newtonian FL = 1200 mm	f/10 8-inch SCT FL = 2000 mm	f/10 10-inch SCT FL = 2500 mm	f/10 12-inch SCT FL = 3048
Eyepiece					
5 mm	120 x	240 x	400 x	500 x	609.6 x
8 mm	75 x	150 x	250 x	312.5 x	381 x
17 mm	35.3 x	70.6 x	117.6 x	117.6 x	179 x
26 mm	23 x	46 x	76.9 x	96 x	117 x
40 mm	15 x	30 x	50 x	62.5 x	76.2 x

Fig. 5.4 Magnification table for different telescopes & eyepieces

things to do. One of the most common questions asked during an outreach event is "How much is it being magnified?" A quick glance at the card provides the answer.

The appropriate magnification to use for a telescope will depend on several factors: the object being observed, the quality of the "seeing", and the telescope being used. *Seeing* refers to the effect that atmospheric turbulence has on the quality of the images observed with a telescope. When there is a great deal of turbulence in the upper atmosphere objects do not appear as sharp as when the atmosphere is calm. It becomes harder to observe details of the planet surfaces when the *seeing* is not good.

In general, lower and mid magnifications (shaded cells in Fig. 5.4) are the best to use under most conditions. The lower the magnification the greater amount of sky you also view. I find using a 32 mm or 40 mm eyepiece when viewing the Pleiades star cluster works well with a f/10 telescope. For Jupiter, I will bump up the magnification using either a 17 mm or 12 mm eyepiece. Only on very rare occasions is the *seeing* is so good that I can use an 8 mm or even a 5 mm eyepiece. You will want to test your eyepieces with your telescope under different conditions and make a note of what works best.

5.2 Observatories

Many astronomy clubs, high schools, colleges, and universities have fixed location observatories. A few organizations, like Science Heads Inc., have built mobile observatories. The Science Heads Inc. mobile observatory design is based upon a standard 12-foot utility trailer. Its light enough so it can be towed using a car or light duty truck. Mounted on top is a 6-foot rotating fiberglass dome. Inside is a custom designed pier that passes through a hole cut in the floor to provide a solid footing on the ground below. The pier breaks down into several parts for easy assembly and transportation. The Science Heads mobile observatory is shown in Fig. 5.5.

Observatories are iconic buildings. When people see a dome on top of a building, they immediately know what is inside. Most people have seen pictures of the famous observatories located at Cerro Paranal, Mount Wilson, Palomar Mountain, Mauna Kea, and Kitt Peak. Many modern discoveries can be traced back to the research done at these locations.

Observatories offer many advantages for outreach astronomy events including:

- Stable mounts for telescopes.
- Multiple telescopes can operate simultaneously.
- Short setup times.
- Protection from poor weather, wind, and extraneous light.
- Computers and interactive workstations.
- Support for using monitors to display images and videos.
- Reliable internet connectivity.
- Wall space for posters, charts, and other information.

Fig. 5.5 The Science Heads Inc. Explorer1 mobile observatory

A mobile observatory offers several additional advantages. It can be:

- relocated to be convenient for the intended audience (ex. set up at a school).
- used to transport equipment (telescopes, chairs, tables).
- set up in dark sky locations when desired.
- support outreaches in nearly all-weather conditions.
- display live images from remote telescopes when weather conditions and the time don't favor local use of telescopes.

Regardless of the type of observatory, visiting one is an experience children and adults both can enjoy and learn from. Observatories are more than just a telescope. They are every tool used by astronomers assembled into one place. A telescope is to an astronomer what a test tube is to a chemist. An observatory is to an astronomer what a laboratory is to that chemist.

Visiting an observatory gives children the opportunity to imagine themselves being an astronomer. Imagination and play are recognized by educators as important factors in a child's education. Visiting an observatory can influence a child's perspective of what they can become. Adults can also gain insight into how science is done. They can see, use, and learn about the tools that professional astronomers use in research. An early concept drawing of the observatory's interior is shown in Fig. 5.6

Guests enter the Explorer1 observatory from the rear of the trailer and exit the side door seen in Fig. 5.7. In between, they are invited to listen to short presentations, ask questions, use interactive computer displays, view live images from the telescope and web sites displayed on the large wall mounted monitors. If weather

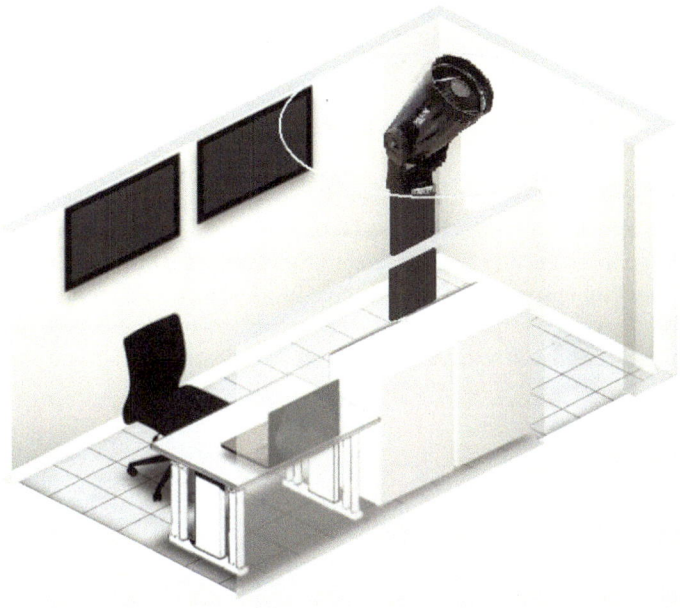

Fig. 5.6 Concept drawing of interior of Explorer1

Fig. 5.7 Interior of Explorer1 setup for daytime solar observations

permits guests can view through the telescope(s) mounted on the central pier. Figure 5.7 shows a typical solar viewing set up with an h-alpha telescope on a Skywatcher solar tracking mount, and a color imager. The live image of the Sun is displayed on the forward monitor.

During cloudy or inclement weather, when local use of the telescope is not possible, observatory operators can connect to remote telescopes via the internet and browse and display live telescopic images. As described in the next section there are a variety of remote telescopes that can be used in this fashion.

Several people can enter the Explorer1 observatory at a time. Science Heads Inc. typically assemblies a group of 7–12 people inside the observatory before starting a 15–20-minute program. Our observatory programs incorporate a variety of presentation materials and software depending on the topics covered and the audience involved.

5.3   Remotely Controlled Telescopes

If you have access to a computer and the internet you can be using a remotely controlled telescope for an outreach event. Even people who have little experience using telescopes can sign up for a subscription to a variety of services that offer these resources.

In most cases remotely controlled telescopes are larger and more capable than the average amateur class telescope. They offer large apertures, stable mounts, high quality imagers, automated filters, and expert on-site personnel for management and maintenance. They are also located at sites that offer better viewing than the urban and suburban locations where outreach events typically take place.

Most remote observatories are built on mountain tops or in locations that offer very dark skies and no light pollution. Using these telescopes, an Outreach Astronomer can count on being able to share amazing images of galaxies, galaxy clusters, nebulas, star clusters, dark regions of space, and star fields. Even viewing planets often surpasses what can be done locally.

There are many observatories that offer subscriptions or pay as you go services. Most of these services will send you the images, if the session was successful, via email or provide a link to download the files. Some services such as iTelescope (https://www.itelescope.net/) and *Slooh* (https://www.slooh.com/) incorporate user interfaces that offer automated processing functions for the images. Most of these services are set up to support research and not necessarily outreach activities.

The *Slooh* service uniquely provides live views of images through their user interface. The live image appears on the screen as the photons are being collected at the telescope's imager. This feature is very useful during outreach events. This live view can be used as a replacement for local telescope viewing during bad weather or cloudy conditions. And because they offer telescopes located on the Canary Islands and in the UAE – images can come from locations where it is night during daylight hours in North America. The Slooh telescopes in Chile offer views of the southern hemisphere. North American users can view objects not observable in local skies.

Using *Slooh*, an Outreach Astronomer can schedule imaging to occur during an event or they can jump in on sessions scheduled by other subscribers and watch as real time images appear. The service supports downloading images in both JPEG and FITS formats. Outreach participants can take their favorite images home with them. *Slooh* offers several levels of commercial subscriptions and special access for educators.

Using remotely controlled telescopes is also a better demonstration of how professional astronomer's work. While some researchers do travel to Cerro Pranal, Mauna Kea, Palomar, and elsewhere – most astronomers work from their office far removed from the telescope that they use. Astronomers analyze, compare, and process image files that were collected the night before or over long periods of time. They employ specialized image processing software, spectroscopy software, and large databases, as they investigate a phenomena or object of interest.

All of this can be better simulated and demonstrated in an observatory. Just as a telescope is a tool – the observatory itself, the computers inside, and the software employed are also tools that make modern scientific research possible. That is part of the message that guests touring an observatory receive.

Using remote telescopes for outreach events offers advantages over using locally setup telescopes:

- Time zone shifting. View nighttime objects during local day time.
- Image storage and processing.
- Easily share images that are downloaded.
- Large aperture telescopes and advanced mounts are available.
- A variety of telescopes offering a variety of focal lengths are available.
- Zero set up time.
- Schedule targets at specific times during the outreach event.
- Encourages guests to do explore astronomy on their own.
- Supports citizen science projects.
- Some offer subject matter expert narrations.

There are a number of sites that offer remote telescope services including those listed in Fig. 5.8:

When incorporating remote telescope usage into an outreach event, it helps to document a schedule. Sometimes observatories get clouded out and having alternatives can help. Listing options you have can help keep the outreach event going. Science Heads Inc. often schedules multiple objects from different observatories

SERVICE	LINK
Slooh (Chile, The Canary Islands, UAE)	https://www.slooh.com/
iTelescope (Australia, Spain, Chile, USA)	https://iTelescope.net/
Remote Observatories (Arizona USA)	https://sierra-remote.com/
Sierra Night Skies Network (member locations)	https://nightskiesnetwork.com/

Fig. 5.8 Remote telescope services

OCASA Event 4/21/2023

SLOOH SCHEDULE

Time (PDT) Tele	Object	Alt		Notes
5:05 PM Chile 2	M42 Orion Nebula			Dist 1,344 LY
5:15 PM Chile 1	Running Man Nebula (NGC 1977)			Ref Neb Dist 1,500 LY
5:30 PM Chile 2	S Galaxy (NGC 2903)			Spiral Dist 30.4 MLY
5:50 PM Chile 1	LMC			Dist 163 KLY 20 b stars
6:05 PM Chile 1	Proxima Centauri (Alph Cent C)			Exo Pl disc 1/2020 Dist 4.2 LY
6:55 PM Canary 2	Dumbbell Nebula (M27)			Plan Neb, Dist 1.2 KLY
7:00 PM Canary 1	Pinwheel Gal M101	Canary 2	Lagoon Nebula	Dist 21 MLY / Emiss Neb 4,100 LY
6:35 PM Canary 2	Eagle Nebula (M16)			Emiss Neb, Dist 5,700 LY
6:55 PM Chile 2	LMC			Dist 163 KLY 20 b stars
7:00 PM Chile 1	Sombrero Galaxy (M104)			Dist 31 MLY
7:05 PM Chile 2	M42 Orion Nebula			Dist 1,344 LY
7:15 PM Chile 1	Sombrero Galaxy (M104)			Dist 31 MLY
7:20 PM Chile 2	Tarantula Nebula (NGC 2070)			Dist 160 KLY
7:30 PM Canary 1	Eagle Nebula (M16)			Emiss Neb, Dist 5,700 LY

Fig. 5.9 Example of a remote viewing schedule

and telescopes just in case an imaging session doesn't occur as planned as the example in Fig. 5.9 shows.

5.4   Solar Viewing

Sharing views of the Sun at outreach events has grown in popularity in recent years. The introduction of affordable hydrogen alpha (h-alpha) telescopes has generally increased interest in helio-physics. The start of a new solar cycle and the resultant increase in solar activity has also helped bring more people into this aspect of astronomy. December 2019 brought in solar cycle 25 and the years since have seen a dramatic increase in sunspot activity. When you share a view of the Sun your guests are now likely to see numerous sunspots, sunspot groups, prominences, plages, and filaments. All that is needed is the right equipment.

5.4.1   Solar Filters

Almost any standard telescope can be turned into a solar telescope by adding a solar filter in front of its objective lens or mirror. Solar filters allow viewing features on the Sun's photosphere, mostly sunspots, by blocking almost all light entering the telescope. Filters are available for different apertures and are made of various materials. To ensure the safety of the viewer, solar filters need to be snuggly and firmly attached to the front of the telescope.

Fig. 5.10 Comparison of solar filters and image color

MATERIAL	IMAGE COLOR
Glass	Yellow Orange
Aluminized Mylar	Blue White
Black Polymer	Orange

Fig. 5.11 Aluminized mylar solar filters

There are three basic types of solar filters commercially available based upon the filter material used. Each type of material produces an image with a slightly different hue as listed in Fig. 5.10. Filters come in various sizes to fit different telescope apertures as shown in Fig. 5.11. Since the Sun is thought of as being yellow, some guests may question seeing the orange or blue hues produced by some filters.

Safety must always be the main concern when viewing the Sun. Solar filters should always be inspected before use. Mylar filters can develop pin holes which can let in dangerous amounts of light. Refer to and follow the manufacturer's instructions for filter maintenance and repair. It is important to take the time to properly explain safety procedures with your guests. Children, will likely not understand the significance and may try looking at the Sun on their own. I rely on a memorized safety talk, so I don't forget important details:

> Never look directly at the Sun. It can damage your eyes. Never look at the Sun with a telescope, binoculars, or even a magnifying glass. It can cause blindness in an instant and it won't be reversable. Only look at the Sun when you are with an adult that is using a specialized telescope, like the one we are using now.

5.4.2 Hydrogen Alpha Telescopes (h-alpha)

In recent years at least three manufacturers (Meade, Lunt, and Daystar) have introduced affordable h-alpha telescopes and accessories. H-alpha telescopes display features on the Sun's chromosphere – the layer between the photosphere and corona. The chromosphere is a region where prominences can be typically seen. It is a fascinating region to observe. A well-tuned h-alpha telescope can show sunspots, faculae, granulation, prominences, plages, and filaments. The image at this wavelength is similar to that for the time-lapse videos often posted on the internet from space based solar telescopes. Many event guests will have seen videos of prominences shooting out into space and will enjoy seeing one with their own eyes.

H-alpha telescopes are offered in a variety of apertures and filter configurations. The filter on an h-hydrogen telescope is called an ètalon, also known as a Fabry-Pèrot interferometer after the names of its inventors. The design incorporates an optical cavity positioned between two mirrored surfaces. Light can only pass through the cavity when it is in resonance with the volume of the void. Tuning the filter adjusts this volume allowing only specific wavelengths of light to pass. For visual use, the ètalon is typically tuned to wavelengths of either 656.3 nm or 396.9 nm. These wavelengths are associated with specific electron transition states of hydrogen and calcium-II respectively.

H-alpha telescopes are available with single or dual ètalons and are referred to as either single or double stack configurations. A double stack offers greater contrast and more detail but also produces a dimmer image since less light reaches the eye. A single stack telescope with a filter bandpass of 0.7 to 0.5 is usually more than adequate for visual use at an outreach event.

5.4.3   Solar Tracking Mounts

Using a solar telescope on a manual or general-purpose computerized mount can pose challenges for the Outreach Astronomer. Older computerized mounts did not include the Sun in their object databases. This was intentional by the manufacturers since looking at the Sun is dangerous. Making it easy could expose the manufacturers to increased liability. Aligning a computerized mount during the day can also be problematic. Most mount controllers require that the operator align the mount using at least one star in the sky (other than the Sun). Getting a computerized mount to accurately track the Sun can therefore require some work arounds.

One technique that has been tried with some success is to sync the telescope controller on Mercury. This planet is always close to the Sun and presents a good starting point. After synching on Mercury, the operator can then point the telescope at the Sun. Using this method tracking the Sun may be possible for an hour or possibly longer. Figure 5.12 shows a Coronado Solar Max II h-alpha telescope on an older style mount.

Some manufacturers of newer computerized mounts have added the Sun to their databases. By leveling the telescope, pointing it north, and syncing on the Sun, the mount should track the Sun reliably throughout the day. Refer to your manufacturer's instructions for the proper procedure.

One manufacturer, Sky-Watcher, has introduced a tracking mount specifically designed for solar telescopes. The Solarquest Heliofind mount (https://www.sky-watcherusa.com) automatically locates the Sun and tracks it with remarkable

Fig. 5.12 H-alpha
telescope on an older
computerized mount

Fig. 5.13 The Sky-
Watcher Solarquest mount

precision. It is an excellent platform for solar telescopes weighing less than 12 lbs.
The Solarquest's built-in camera reliably locates the Sun usually with no problems.
Science Heads Inc. uses this mount (see Fig. 5.13) with a Meade Coronado SolarMax
II h-alpha telescope for its daytime outreach events. The combination provides a
stunning view of the Sun's chromosphere for hours at a time.

5.5 Electronically Assisted Astronomy

Electronically Assisted Astronomy (EAA) refers to the use of telescopes and imagers used to display objects on computers and monitors. It has grown in popularity as more people venture into astrophotography (AP). EAA and AP often get confused as discussions turn to related subjects like image processing, binning, stacking, etc.

Simple EEA setups can be particularly useful when working with audiences who otherwise would not be able to look through a telescope. This includes aging adults, very young children, and individuals who do not have the agility or mobility to use a telescope. It can also be appropriate when presenting to large groups of people.

Implementing a simple EAA setup for an outreach need not be expensive nor difficult. The minimum required hardware is readily available and affordable:

- Telescope with a stable mount, tripod, or pier.
- Low-cost imager with USB connection.
- Computer with USB and HDMI ports.
- Imaging software.
- TV monitor with HDMI connection.
- HDMI and USB cables.
- Power source for computer and TV monitors.

Optional items that may also be useful with EAA include:

- Flip mirror.
- Eyepiece with reticle.
- Parfocal ring.
- TV monitor stand(s).
- Extended length USB and HDMI cables.
- HDMI splitter.
- Second TV monitor with HDMI.

Virtually any telescope can be used for EAA at an outreach. The main requirement is that the focuser has sufficient travel to achieve focus on the imager's CCD or CMOS chip. Science Heads Inc. uses black & white and color imagers manufactured by ZWO.

There are many choices for imagers available for under $ 500 US. All of these could be used for outreach EAA. Manufacturers include:

Company	Link
Celestron	https://www.celestron.com/
Mallincam	https://www.mallincam.net/
Meade	https://www.meade.com/
Orion	https://www.telescope.com/
QHY	https://www.qhyccd.com/
Revolution imager	https://revolutionimager.com/
ZWO	https://astronomy-imaging-camera.com/

Fig. 5.14 Imager, Reticle Eyepiece, Flip mirror, and Parfocal Ring

The Revolution Imager is unique in that it includes a controller with built in processing software and a small 7-inch monitor that can be mounted on a telescope. The other imagers include a USB port or either come with compatible software or can be used with readily available processing software including:

Software	Link
Jocular	https://transpy.eu.pythonanywhere.com/jocular/
SharpCap	https://www.sharpcap.co.uk/

Various telescope accessories can be employed with imagers including those shown in Fig. 5.14.

Depending on the imager model used, industry standard Astronomy Common Object Model (ASCOM) drivers may be required. ASCOM drivers are available at https://ascom-standards.org/Downloads/Index.htm.

An alternative to buying separate components is to purchase a telescope designed specifically for EAA. Very recently some manufacturers have introduced dedicated purpose combined telescope/imaging systems designed specifically for Electronically Assisted Astronomy. These systems typically connect to apps running on smart phones. Since they are designed only for EAA, eyepieces cannot be used with these telescopes. Some industry experts believe this technology represents the future of amateur astronomy.

Two manufacturers offering systems of this type are:

Company	Offering	Link
Unistellar	(eVscope and eQuinox)	https://shop.unistellar.com/
ZWO	(Seestar)	https://astronomy-imaging-camera.com/

When using EAA for outreach it is best to keep the setup as simple as possible and practice using it in advance of an event. Employing EAA adds more details to your set up. It involves more than just assembling a telescope, aligning the mount, and sloohing to an object.

- The focus for an eyepiece and the imager will be significantly different.
- Using processing software to display a useful image can also take some time. It involves finding the right parameters for the object such as exposure, gain, and color balance.
- There will be more cables and connections to make.

Using EAA for outreach can add 45 minutes to an hour to a typical telescope setup. Standardizing the equipment, the software, the settings, and practicing beforehand can help reduce set up time.

If you are going through the extra effort of using EAA, consider the proper placement of monitors for your audience. During the day outdoor monitors can get washed out by ambient light. Placing them under easy-ups (aka popup shades) or under the shade of a tree improves contrast. Using monitors with the highest lumens specs possible can also help. Generally, monitors for exterior use are rated around 1200 NITS (aka candelas).

Using multiple monitors gives your audience options to position themselves so they can better see the images. This is important with larger crowds. Consider cordoning off an area in front of the monitors for kids to sit, allowing the adults to stand behind.

Science Heads Inc. has been very successful providing satisfying experiences to many types and sizes of audiences. Presenting to groups of 30–50 people is commonly done. By employing EAA you can accommodate audience members who are limited by age, agility, or mobility. Not being able to climb steps, see through a telescope, or access an observatory need not be a hindrance to participation in your outreach events. Two examples of EAA setups are shown in Figs 5.15 and 5.16.

5.6   Resources for Blind and Limited Vision Audiences

Chapter 3 described a personal experience I had interacting with a blind individual at the California Science Center. Fortunately, at the time, I had in my possession the tools needed to make the interaction rewarding and enlightening for my guest. A small model, a tile sample, and willingness to use language expressively was all that was needed. But employing these techniques my guest was able to use his hands and mind to "see" the Endeavour space shuttle.

Models and tools are also available for astronomy outreach, and many are free. With funding provided by NASA, several premier universities and institutions together created the *Universe of Learning* web site just for this purpose (https://www.universe-of-learning.org/contents/products/touchable-universe-in-a-box).

Fig. 5.15 An EAA setup for solar viewing

Fig. 5.16 A dual monitor EAA presentation

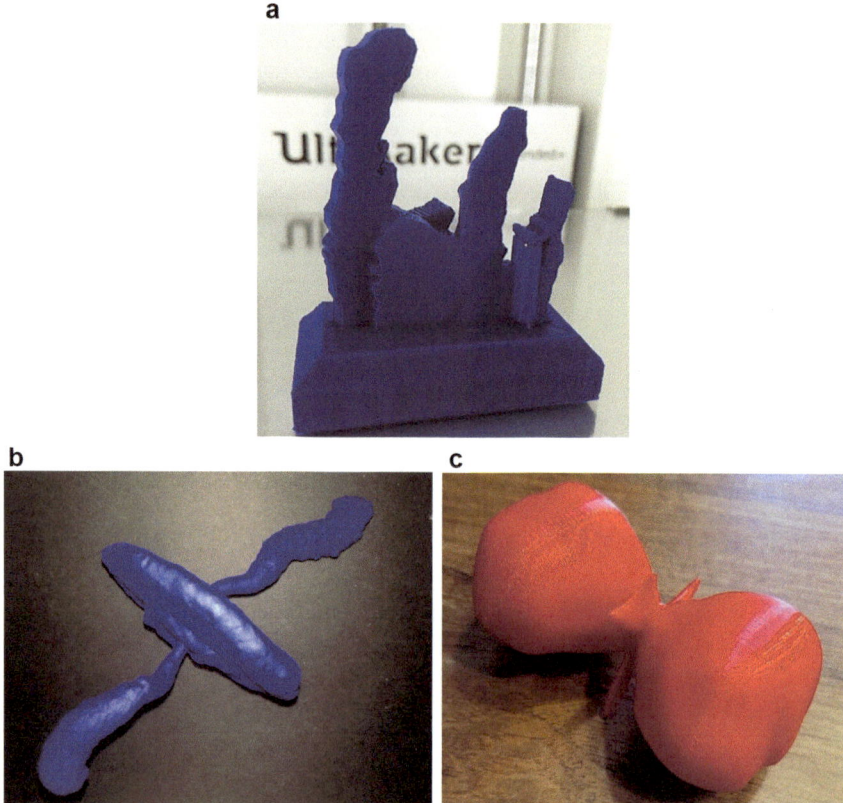

Fig. 5.17 (**a**) The Pillars of Creation. (**b**) The Crab Nebula. (**c**) Eta Carinae. (Credit: NASA)

Educators are welcome to download 3D print files, tactile plates, Braille/tactile posters & cards, visual descriptions, and data sonification files.

Imagine trying to describe a supernova remnant to a blind individual. Without being able to see the nebula it would be very difficult to imagine its shape and contours. But by downloading and printing a free STL or Makerbot file and printing it on a 3D printer the Outreach Astronomer can offer the person a model to hold and feel in their hands.

The *Universe of Learning* web site provides links for 3D print files of several objects including a supernova remnant, a binary star system, and various nebulas. All that is needed is a 3D printer and a little time to produce the models shown in Figs 5.17a, 5.17b, and 5.17c.

Braile books are another way to support blind and limited vision guests. Books of this type provide not just 2 dimensional models of astronomy objects but also detailed descriptions and explanations. There are several publishers and sources for braille astronomy books including those listed below, an example of which is shown in Fig. 5.18.

Publisher	Link
Haptically Speaking	https://hapticallyspeaking.com/
National Braille Press	https://www.nbp.org/
You Can Do Astronomy, LLC	https://youcandoastronomy.com/

Fig. 5.18 Braille page about the planet Jupiter. (Credit: N. Grice)

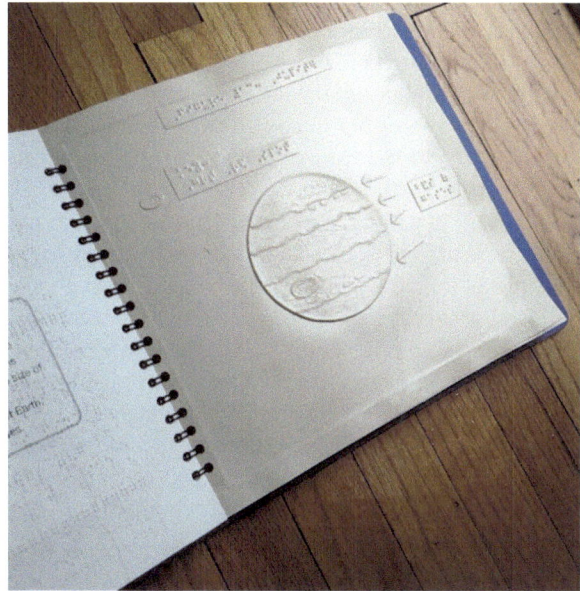

5.7   Planning Software

There are two types of software available that are useful for the planning and execution of astronomy outreach events. They fall into the categories of Planetarium and Simulation/Visualization software.

Planetarium software is generally designed to calculate and display the positions of objects on specified dates and times. The user interface typically renders a map of the sky with the objects shown in their relative positions. This can be very useful for planning purposes. An astronomer can better understand where to find planets, nebulas, star clusters, and other objects at the current time or some time in the future. Previous dates and times can also be input to display what the sky looked like in the past.

Many planetarium software packages offer advanced features such as controlling a telescope mount, generating observing lists, keeping logs of observations, listing best viewing opportunities, and even rendering a full dome image for a planetarium. The perspective presented by planetarium software is almost always that of an Earth based observer. It shows what a person standing on the surface of Earth would see.

Simulation/Visualization software differs in that the perspective of the observer can be anywhere. The perspective can be from Earth, Mars, on the moon Europa looking at Jupiter or back at the Earth; it can be hovering above the solar system; or even flying through space on a specified trajectory. There are potentially an unlimited number of perspectives that can be rendered.

The functionality of these two categories of software sometimes overlaps. Both simulation/visualization and planetarium software may use actual images to render objects. Pictures captured by Hubble and other space-based telescopes are often incorporated into both types of software.

But while planetarium software often uses static images, some simulation software incorporates 3D images. The *NASA Eyes on the Solar System* simulator for example shows not just a static image of Jupiter but an image that it can rotate so the correct side is facing the viewer from their perspective on the specified date and time. This is particularly useful if you want to know where the Great Red Spot (GRS) is currently located. If it's not on your side of Jupiter currently, you can rotate the planet to find it and determine what date and time it will become visible.

Both categories of software can be very useful when preparing and giving presentations. Capturing a star chart for a slide is very easy using planetarium software. If you want to include a visualization of what the night sky looks like from Mars, it's easy to do using a simulation/visualization package.

Some simulation/visualization packages also provide functionality for recording and playing scripts. This can be useful to generate an animation of the motions of planets, moons, the sky, and more. You could even create a flyby tour of the solar system and incorporate that into a presentation. The number of possibilities using this type of software is truly endless.

Figure 5.19 lists some of the more popular free planetarium software. Besides these there are many other titles available with a paid license or subscription. Packages are supported for all popular platforms (Windows, MacOS, iOS, Android, Linux).

Figure 5.20 lists free simulation/visualization software for astronomy. Example screens from NASA's Eyes on the Solar System are shown in Figs 5.21 and 5.22.

PLANETARIUM SOFTWARE	SOURCE	SUPPORTED OPERATING SYSTEM
Cartes du Ciel	https://www.ap-i.net/skychart/en/start	Linux, macOS, Windows
HNSKY	https://www.hnsky.org/software.htm	Linux, Rasberry Pi, Windows
KStars	https://kstars.kde.org/	Linux, macOS, Windows
Stellarium	http://stellarium.org/	Android, Linux, macOS, Windows, Web

Fig. 5.19 Free planetarium software

SIMULATION SOFTWARE	SOURCE	SUPPORTED OPERATING SYSTEM
Celestia	https://celestiaproject.space/	AmigaOS, Android, iOS, macOS, Linux, Windows
Eyes on the Solar System	https://eyes.nasa.gov/apps/solar/#/home	Web only
Eyes on Asteroids	https://eyes.nasa.gov/apps/asteroids/	Web only
Eyes on the Earth	https://eyes.nasa.gov/apps/earth/#/	Web only
Eyes on Exoplanets	https://eyes.nasa.gov/apps/exo	Web only
Sky Tonight	App Store, Google Play	iPhone, Android
Star Walk	App Store, Google Play	iPhone, Android
Night Shift	Google Play	Android
OpenSpace	https://www.openspaceproject.com/	MacOS, Windows

Fig. 5.20 Free simulation/visualization software

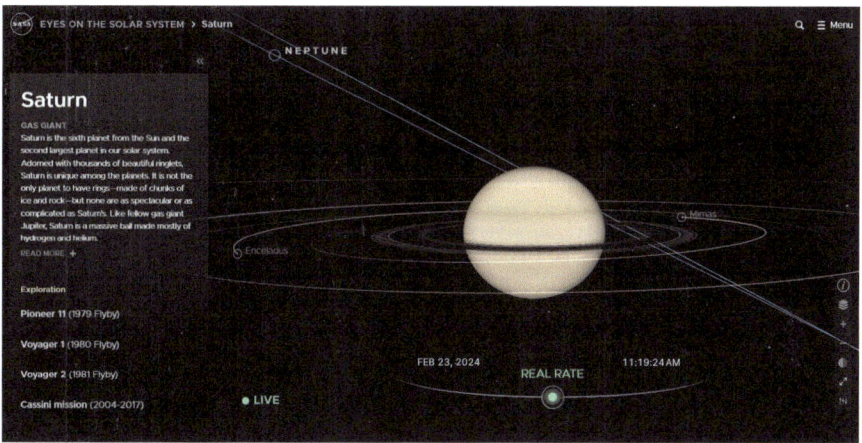

Fig. 5.21 Saturn rendered with NASA/JPL Eyes on the Solar System. Credit: NASA

5.8 On-Line Events

Not all astronomy outreach events are in person. The advent of internet- based meeting and webinar services has opened the door to new ways to reach audiences. With these services it is now possible to interact and present to audiences that are widely dispersed, even across the globe. During the pandemic years of 2020 through 2023, many people were introduced to this functionality. Workers performed their job duties from home. Students attended classes remotely.

On-line meeting and webinar software has been around for many years. For decades businesspeople relied on these services as an alternative to travel.

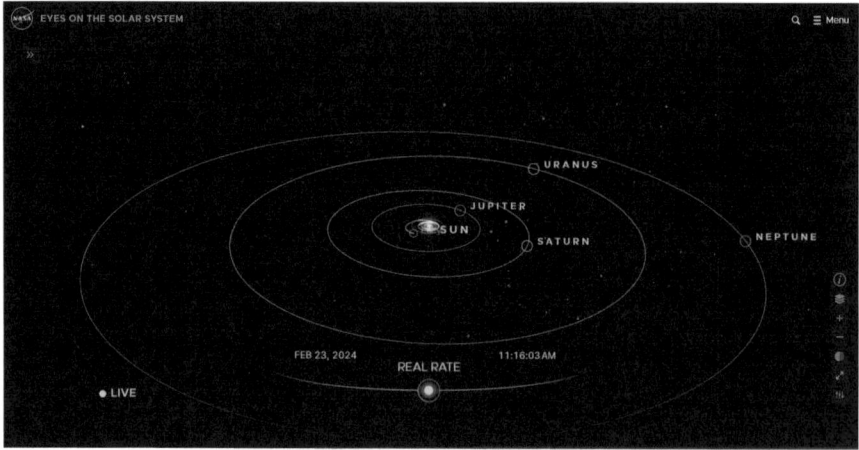

Fig. 5.22   The solar system rendered with NASA/JPL Eyes on the Solar System. Credit: NASA

There are many commonalities between on-line meetings, webinars, and live streaming sessions. There are also many differences. All support the sharing of a presenter's screen and applications with attendees. All provide mechanisms for the attendees to interact during the session such chat and question & answer communications. On-line meeting software allows the greatest amount of audience participation. The software is modeled on in-person meetings. Everyone gets the ability to talk via voice over IP (VOIP), share screens (one at a time), and upload documents. Some meeting software also supports white board functionality and collaborative editing of files. On-line meetings are based on a many-to-many architecture.

Webinar software on the other hand is modeled on a lecture model. This is a one-to-many architecture. One, or a handful of presenters, share their screens and give a presentation to the entire audience. The audience does not typically have the same presentation rights. They are not able to share or talk. Instead, they can type in questions in a chat or question dialog box and then wait for a response. Some software of this type includes a raise hand function so audience members can get the presenter's attention.

The webinar architecture works well for an astronomy outreach event. Presenters are given full control over content, and they can allow input and comments at appropriate times. On-line meetings can also work well for outreaches if the audience is small, and a lot of interaction is appropriate. Both types of services typically support recording the sessions. This is handy for those who were not able to attend the meeting or webinar. The recordings can be uploaded to a web site or streaming site like YouTube.

Live streaming is the most audience limiting architecture of the three types of on-line services. Its model is that of a TV broadcast although it can support some limited feedback. The live streaming starts and ends when the presenters decide.

The audience typically has no ability to directly communicate with other audience members. Some live streaming platforms limit the length of the streamed program.

During the pandemic Science Heads Inc. held dozens of astronomy outreach webinars. It also hosted on-line meetings for schools. The webinars were posted to the Science Heads Inc. YouTube channel (https://www.YouTube.com/@science-heads3077) and remain there for anyone to stream on demand. These webinars covered a wide variety of astronomy and space related topics including:

- Discovering Asteroids & Comets.
- Measuring the Universe – a Herstory for Womens History Month.
- Flying the Space Shuttle – a Black History Month Special.
- Introduction to Using a Telescope.
- Constellations and Ancient Stories.
- Our Dynamic Sun.
- The Life of Stars.
- The Gas Giants.

The school events Science Heads Inc. hosted replaced in-person events that were planned but could not be held because of the pandemic. Even though we were not in person with telescopes and hands-on activities the students joined talks and had an opportunity to interact with local graduate students.

Figure 5.23 shows the program for one of these events. The middle school students who attended were able to choose the subjects that interested them and attend as many sessions as desired.

Listed in Fig. 5.24 are some of the better-known on-line meeting and webinar services. Most of these services require a subscription.

5.8.1 Live Streaming

Besides on-line meetings and webinars, the Outreach Astronomer has the option of live streaming an event. Many popular social media platforms make this possible. Each platform differs in the set up required and ways to market the content. In general, it is best to use the platform on which you do most of your social media posting. This ensures that your followers and subscribers are aware of your event.

Live streaming content can be created in many ways. It can be as simple as holding a smart phone in front of you. Or it can involve many people, many sources of video, images, audio, even previously recorded videos.

Not much preparation is needed for the simplest approach beyond an outline and a few practice runs. But when you involve many people and plan to use many sources it becomes necessary to employ mixing software designed for the purpose. One of the most popular programs for this is a free package called OBS Studio (https://obsproject.com/). OBS Studio is a professional grade production mixing application that allows you to configure and mix several sources (ex. multiple cameras for different angles or multiple people) into one video and audio stream. OBS

Ethan Allen Elementary School
Virtual Astronomy Night
Wednesday, December 16th, 2020

Hello and welcome to the Ethan Allen Elementary Virtual Astronomy Night. Attend one activity during each session by clicking on the link in the right hand column below. Please join the activity at least 5 minutes before the start time. You may attend Activity 5 & 6 anytime. You may attend as many activities as you like - even attend multiple times. We hope that you enjoy the evening!

SESSION	START TIME	DURATION	SESSION NAME	DESCRIPTION	MATERIALS NEEDED AT HOME	SESSION LINK
1	6:30 PM	15 MINS	Welcome Session	Science Heads volunteers will explain what activities are being offerered during tonights Virtual Astronomy Night and how to participate.	None	Attend Session
2	7:00 PM	20 mins	Map the Moon	Resident artist and NASA Solar System Ambassador Ruth Kurisu will lead students in an exercise to create a map of the lunar surface.	5 oz liquid dishwashing detergent (without degreaser), 5 oz black or blue tempera paint (acrylic paint will not work), small styrofoam or Dixie paper bowl, Straw, White Cardstock, Water.	Attend Session
		20 Mins	Story Time	NASA Solar System Ambassador Cheyenne Smith will read one of her favorite children's book about astronomy.	None	Attend Session
		20 mins	Explore the Solar System	NASA Solar System Ambassador Richard Stember will take the students on a virtual tour of our Solar System using NASA software and images.	None	Attend Session
		30 mins	The Search for Life	NASA Solar System Ambassador Michal Peri, Ph.D. gives a talk about NASA's search for life beyond Earth.	None	Attend Session
3	7:30 PM	20 mins	Map the Moon	Resident artist and NASA Solar System Ambassador Ruth Kurisu will lead students in an exercise to create a map of the lunar surface.	5 oz liquid dishwashing detergent (without degreaser), 5 oz black or blue tempera paint (acrylic paint will not work), small styrofoam or Dixie paper bowl, Straw, White Cardstock, Water.	Attend Session
		20 mins	Story Time	NASA Solar System Ambassador Cheyenne Smith will read one of her favorite children's book about astronomy.	None	Attend Session
		20 mins	Explore the Solar System	NASA Solar System Ambassador Richard Stember will take the students on a virtual tour of our Solar System using NASA software and images.	None	Attend Session
4	8:00 PM	20 mins	Map the Moon	Resident artist and NASA Solar System Ambassador Ruth Kurisu will lead students in an exercise to create a map of the lunar surface.	5 oz liquid dishwashing detergent (without degreaser), 5 oz black or blue tempera paint (acrylic paint will not work), small styrofoam or Dixie paper bowl, Straw, White Cardstock, Water.	Attend Session
		20 mins	Story Time	NASA Solar System Ambassador Cheyenne Smith will read one of her favorite children's book about astronomy.	None	Attend Session
		20 mins	Explore the Solar System	NASA Solar System Ambassador Richard Stember will take the students on a virtual tour of our Solar System using NASA software and images.	None	Attend Session
		30 mins	The Search for Life	NASA Solar System Ambassador Michal Peri, Ph.D. gives a talk about NASA's search for life beyond Earth.	None	Attend Session
5	ANYTIME	40 mins	Constellations and Ancient Stories	Join Science Heads volunteers and NASA Solar System Ambassadors Cheyenne Smith and Richard Stember in this pre-recorded video introduction to finding constellations. Learn about simple tools and tips that help. Our hosts also share ancient stories about constellations from Native American, Mayan, and ancient Greek civilizations.	None	Play Video
6	ANYTIME	N/A	Virtual Art Wall	Email us a picture of you holding your art work and we will post it on our Virtual Art Wall. Click on the links to the right to submit and view your artwork.	Email Your Picture	View Art Wall
7	8:45 PM	15 mins	Wrap Up Session	Join us in this final session to express gratitude to your teachers, our volunteers and all of the students who participated. A special virtual gift will be provided to those in attendance.	Computer printer	Attend Session

Science Heads Inc. is a 501(c)(3) non-profit on a mission to increase science literacy. Visit www.ScienceHeads.org to sign up for a monthly newsletter and learn more. Donations help us reach more students and are greatly appreciated. Thank you!

Fig. 5.23 Example of a virtual astronomy outreach program

SERVICE PROVIDER	LINK
Adobe Connect	https://www.adobe.com/products/adobeconnect/
GotoMeeting	https://www.goto.com/meeting/
GotoWebinar	https://www.goto.com/webinar/
Microsoft Teams	https://www.microsoft.com/en-us/microsoft-teams/
Vimeo	https://vimeo.com/features/webinar/
Webex	https://www.webex.com/
Zoom	https://zoom.us/

Fig. 5.24 Popular meeting and webinar service providers

LIVE STREAMING SERVICE	LINK
Discord	https://discord.com/
Facebook Live	https://www.facebook.com/formedia/tools/facebook-live
YouTube Live	https://www.youtube.com/live
TikTok	https://www.tiktok.com/live
Vimeo	https://livestream.com/
X (Twitter)	https://business.twitter.com/en/blog/go-live-on-twitter.html

Fig. 5.25 Popular live streaming service providers

Studio provides functions for merging graphic overlays, static images, and animations. It also automates the connections to popular live streaming platforms including those listed in Fig. 5.25.

While OBS Studio can offer you high production value it's important to remember that TV shows are not produced by one person. It takes a team to create quality. Consider assembling an appropriate group of people to help produce your event. You may need several individuals performing tasks such as:

- Script writer.
- Graphics designer.
- Audio and Video Tech.
- OBS Studio operator/producer.
- Camera operator(s).
- Lighting.
- On air personality(s).
- Music producer/director.
- Post production producer/editors.
- Web site developer.
- Social media coordinator.
- PR coordinator.

Chapter 6
Getting Started

There is no better way to learn how to do outreach events than working with experienced people. That is how I got started and learned how to effectively organize and conduct events. Even after you have participated in many events there will always be new things to learn.

Fortunately, there are many organizations willing to help. Several non-profit organizations and even some government agencies offer free lectures, seminars, training videos, and materials. Signs, posters, and slides are available on-line to download and use in your events. And there are many astronomy clubs around the world that sponsor on-going outreach activities that need volunteers.

6.1 Partnering with Organizations

Joining a local astronomy club is a great way to learn. A list of U.S. astronomy clubs and events they are hosting is available on NASA's Night Sky Network website (https://nightsky.jpl.nasa.gov/clubs-and-events.cfm).

Another source for locating clubs is the Astronomical League (AL) (https://www.astroleague.org/). The AL is a national umbrella organization whose membership includes over 300 clubs and associations across the United States.

And my own organization, Science Heads Inc. (http://www.scienceheads.org/) is recruiting new volunteers and looking to form new chapters. Science Heads Inc. hosts not only astronomy outreaches, but also events that focus on all of the STEM subjects (science, technology, engineering, and math).

© The Author(s), under exclusive license to Springer Nature Switzerland AG 2024   109
R. Stember, *Share the Universe*, The Patrick Moore Practical Astronomy Series,
https://doi.org/10.1007/978-3-031-53495-9_6

6.1.1 Astronomy Clubs

By joining an existing outreach program you will work alongside many knowledge-able and capable people. You will also meet many likeminded people interested in informal education. Joining a club may not always be possible. Not all regions are served by astronomy clubs and not all clubs have outreach programs.

In these cases, the Outreach Astronomer has two choices: convince your club to start an outreach program and probably end up running it, or form your own organization focused on doing outreach events.

Most astronomy clubs have legal status in the state in which they operate. They are typically incorporated entities that have been granted non-profit status. This status is necessary to avoid having to pay taxes on money that is raised or donated to the organization. Gaining non-profit recognition from the Internal Revenue Service (IRS) provides the additional benefit of being able to accept tax free donations. People who donate to your organization under many circumstances can write off the donation on their tax returns. This encourages donations that benefit the community and incentivizes donors by effectively reducing their tax burden. Also, it should be noted that foundations and grant providing agencies often only give funds to incorporated non-profits. They rarely give grants to individuals.

Having legal status as a corporation gives an organization and its outreach program benefits beyond tax free status. A properly run corporation shields its board members, officers, and directors from legal liability. The corporation can also protect its volunteers.

The organization should and may be required to purchase policies for:

- Directors and Officers Insurance (D&O).
- General Liability Insurance (GL).
- Workers Compensation Insurance.
- Vehicle and Inland Marine Insurance.

For all these reasons it's often desirable to join an existing outreach program or start one under the auspices of an existing non-profit organization.

6.1.2 Academic Outreach Programs

Many astronomy departments at universities, colleges, and observatories sponsor ongoing outreach programs. Volunteering to help one of these programs may offer exciting opportunities such as meeting and working with researchers and professional educators.

Figs. 6.1 and 6.2 list several universities and observatories that currently offer outreach programs. Details about each program and information for volunteers can be found at the respective links.

OBSERVATORY	LOCATION	LINK
Lowell Observatory	Flagstaff, AZ	https://lowell.edu/support/volunteer/
Kitt Peak Observatory	Southwest of Tucson, AZ	https://kpno.noirlab.edu/support/
Griffith Park Observatory	Los Angeles, CA	https://griffithobservatory.org/support/volunteer/
Mount Wilson Observatory	Los Angeles, CA	https://www.mtwilson.edu/volunteer/
Palomar Observatory	San Diego, CA	https://sites.astro.caltech.edu/palomar/community/docents.html
McDonald Observatory	Austin, TX	https://mcdonaldobservatory.org/volunteer/
Goldendale Observatory	Goldendale, WA	https://www.parks.wa.gov/262/Volunteer-Program/
Yerkes Observatory	Williams Bay, WI	https://yerkesobservatory.org/join/volunteer/

Fig. 6.1 Volunteer opportunities at professional observatories

UNIV.	LOCATION	LINK
Univ. of Arizona	Tucson, AZ	https://www.lpl.arizona.edu/outreach
Caltech University	Pasadena, CA	https://astro.caltech.edu/outreach
Univ. of CA, Irvine	Irvine, CA	https://www.physics.uci.edu/outreach
UCLA	Los Angeles, CA	https://www.astro.ucla.edu/astronomy-live.html
Univ. of Iowa	Iowa City, IA	https://physics.uiowa.edu/outreach
Valparaiso Univ.	Valparaiso, IN	https://www.valpo.edu/physics-astronomy/about/astronomy-outreach/
Hofstra Univ.	Hempstead, NY	https://hofstra.edu/physics-astronomy/observatory.html
Cornell Univ.	Ithica, NY	https://astro.cornell.edu/outreach
Columbia Univ.	New York, NY	http://outreach.astro.columbia.edu/
Ohio State Univ.	Columbus, OH	https://astronomy.osu.edu/outreach
Clemson Univ.	Clemson, SC	https://www.clemson.edu/science/academics/departments/physics/outreach/planetarium.html
Univ. of Tenn.	Knoxville, TN	http://www.phys.utk.edu/trdc/
Rice Univ.	Houston, TX	https://space.rice.edu/outreach.html
Univ. of WA	Seattle, WA	https://astro.washington.edu/jacobsen-observatory

Fig. 6.2 Selected university astronomy outreach programs

6.2   Resources for Training

NASA (National Aeronautics and Space Administration https://www.nasa.gov/) is an independent agency of the United States government and a premier research organization. One of its core missions is to support formal and informal education.

NASA itself funds and organizes many public outreach efforts. It also provides outreach materials and training to educators and the public.

NASA's *Universe of Learning* web site (https://www.universe-of-learning.org/) provides ready to use materials that can be incorporated into your events. It also offers regularly scheduled professional development seminars on recent discoveries and astronomy topics. These seminars are given by prominent scientists, engineers, and researchers who work for NASA.

If you work or volunteer at a museum, school, or non-profit organization, joining NASA's *Museum & Informal Education Alliance* (https://stemgateway.nasa.gov/connects/s/mie-alliance-landing-page) is a great way to stay informed about the latest NASA missions and research. The alliance offers its members free monthly seminars and access to NASA produced videos, slide presentations, literature, and other materials. The alliance's newsletter also provides links and information on NASA grant opportunities and programs.

The NASA CONNECTS website (https://stemgateway.nasa.gov/connects/) is designed to make it easy to find NASA materials that are useful to both formal and informal educators. The site supports collaboration, peer to peer sharing of best practices, and on-line training webinars. Membership is not required, although Museum & Informal Education Alliance members are provided access to collaboration features, training resources, and a monthly newsletter.

NASA/JPL Solar System Ambassador Program

For the past 11 years I have had the honor and privilege to volunteer with the NASA Solar System Ambassador program (https://science.nasa.gov/engage/solar-system-ambassadors/). Solar System Ambassadors (SSAs) receive professional development training provided by NASA engineers, principal investigators, and researchers. Topics include ongoing and planned NASA missions and areas of astronomy

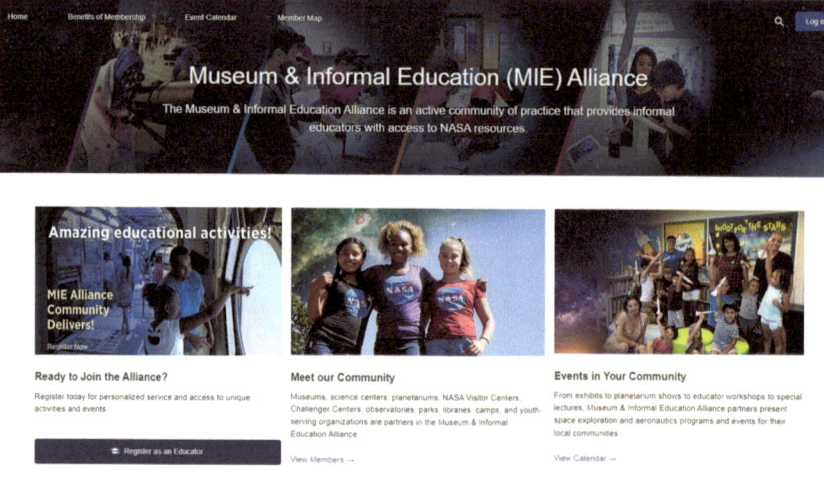

Fig. 6.3 NASA Museum & Informal Education Alliance website. Credit: NASA

research. Over 1200 ambassadors are tasked with sharing this information with their communities. SSA's engage their audiences with various types of public outreaches at schools, libraries, museums, and other public venues.

Over the years I have given numerous presentations about NASA and its missions to space. I have had the pleasure to team up with other SSAs, traveled to NASA sites, and interacted with NASA scientists. The experience is educational, thrilling, and very rewarding.

Hosting events to educate the public about NASA is one of my passions. It also dovetails very nicely with my other passion – hosting astronomy outreach events.

Whether in front of school aged children or grown adults, in a school parking lot or a conference room at a library, or in front of a large audience in an auditorium – the people in attendance at my SSA talks have always been keenly interested in learning about NASA and space exploration. I have found all the techniques described in this book useful in this effort.

I frequently combine Solar System Ambassador events with a star party or a family STEM/STEAM event. Libraries and museums are often very willing to provide the space and help with the planning and coordination required.

Fig. 6.4a Apollo 50th Anniversary presentation poster

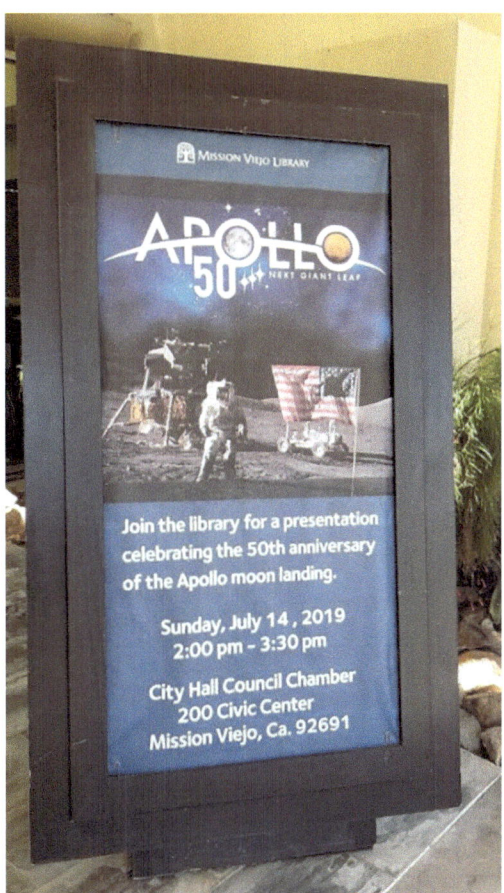

Fig. 6.4b Apollo 50th Anniversary presentation in city council chambers

Combining a talk with an outreach event helps draw a larger audience. My Apollo 50th anniversary talk shown in Figs. 6.4a and 6.4b combined a talk with a star party focused on the Moon. Guests learned about the Apollo 11 mission and were then thrilled to observe Mare Tranquillitatis, where Neil Armstrong and Buzz Aldren landed.

Each year, during the month of September, people who are interested in becoming an SSA can apply via the SSA website. No special expertise is required, only a sincere interest in NASA and a desire to share information with the public. Solar System Ambassadors are volunteers, not employees of NASA, and are not official spokespeople. SSA's are informal communicators who share their knowledge and enthusiasm for the U.S.'s space program.

6.3 Organizing for Outreach

Doing astronomy outreach well can often require a team. While one individual can put on very effective informal education events it often helps to get other people involved. The more people helping the larger the potential impact. There are many ways to find people to help.

6.3.1   Starting a New Organization

If you are starting a new non-profit organization, you may discover that the process of incorporating is not that difficult. It can often be done online at the Secretary of State or Department of Corporations web site for your state government. Searching for "incorporating a business" and your state's name should list the appropriate web pages and forms.

> **Important Note**
> The information provided herein is for informational purposes only. It should not be construed as legal advice. You should always seek legal advice from a licensed attorney.

Registering as a corporation typically requires first writing bylaws for your organization. There are many on-line resources available for this process. Some commercial legal sites provide fill in the blank templates for each state to automate the process.

Statement of information forms and incorporation papers will need to be signed by the initial directors or board members and submitted to the proper authority in your state government.

While the process may not be difficult, it may be best to first consult with a business attorney. Corporations need to be run following specific rules and procedures to maintain their legal status. There are many books and resources on the subject that can help explain the process and requirements.

After your group is recognized by your state, you will need to apply for an Employer Identification Number (EIN) from the IRS. An EIN is comparable to a social security number but for a business. It identifies the organization and is needed for submitting forms to states and federal agencies. EIN's are also required by funding sources.

Once the EIN is obtained then your group can apply for non-profit recognition with the IRS. There are different classifications of this status, but the process is also generally very simple. The IRS recently introduced a simplified form, the 1023-EZ, that groups expecting less than $ 50,000 in annual revenue, can use. (See the IRS page at: https://www.irs.gov/forms-pubs/about-form-1023-ez). From start to finish the above process can often be completed in less than 3 months.

While you are waiting to receive non-profit status, it is a good time to start recruiting volunteers. Give a presentation on what you plan to do at local clubs and organizations. Many of them have gone through the non-profit incorporation process before and, besides being a source of volunteers, they may be able to offer you advice. Find as many interested people as you can. Volunteers are not like employees. Volunteers come and go as their interest and availability allows. The larger the list of volunteers you have the more you will be able to accomplish.

Your organization may be a non-profit, but it still must be run like a business. This means tasking people to do essential things like:

- Keeping the books.
- Answering calls and scheduling events.
- Recruiting volunteers.
- Purchasing necessary materials.
- Maintaining contacts at local schools, organizations, media outlets.
- Web site design and maintenance.
- Marketing and advertising.
- Fund raising.

Insurance

Many venues including schools, parks, libraries, and museums require that groups hosting an event show proof of liability insurance. This typically involves purchasing a General Liability policy with $ 1,000,000 or greater coverage. Fortunately, policies of this type are not expensive for organizations doing public outreach events. Check with a local insurance agent for options that are available.

Some venues also require workers comp insurance if the organization has employees. If you do not have any employees, signing a waiver may be acceptable to the venue.

Other types of insurance may also be needed depending on your organization. Science Heads Inc., for example, purchases Inland Marine Insurance policies for its mobile observatories. This covers replacement costs should the observatories be damaged or destroyed.

6.3.2   Recruiting Volunteers

Outreach events can require many people. Unless you have the funds to hire staff you will likely be looking for volunteers. As described earlier a good source of volunteers are local astronomy clubs. The NASA Night Sky Network and the Astronomical League web sites both list clubs and contact information.

Also consider approaching local high schools and colleges. Many schools have astronomy clubs and students who are eager to help. High schools often require that students volunteer several hours with local non-profits to graduate. This is the perfect opportunity for you to recruit volunteers and help them graduate from school.

There are several websites that can help with recruitment. Science Heads Inc. has relied on *VolunteerMatch.org*, for much of its recruitment efforts. Posting volunteering opportunities is often free for non-profits.

Sites that publish volunteering opportunities:

Website	Link
DoSomething	https://www.dosomething.org/us
Engage	https://engage.pointsoflight.org/
GivePulse	https://learn.givepulse.com/
JustServe	https://www.justserve.org/
VolunteerMatch	https://www.volunteermatch.org/

Finding volunteers is just the first step to adding them to your team. Training, organizing, scheduling, and communicating with them are typical on-going tasks you also need to consider.

There are several on-line services and applications that can be used for this purpose. Most require a subscription. These services include:

Service	Link
Galaxy Digital	https://www.galaxydigital.com/
Volgistics	https://www.volgistics.com/
Volunteer Scheduler	https://volunteerschedulerpro.com/

You may find that using simple spreadsheets works just as well for your organization. The above commercial applications automate communications and provide collaborative environments so volunteers can login and sign up for shifts. For events that are not repeated, Science Heads Inc. use a Googles Doc spreadsheet saved to a shared Google drive location. The spreadsheet is used to keep track of event details and volunteer participation. It also makes it easy to create a mailing list for communications simply by filtering and copying. The headers of this spreadsheet are shown in Fig. 6.5.

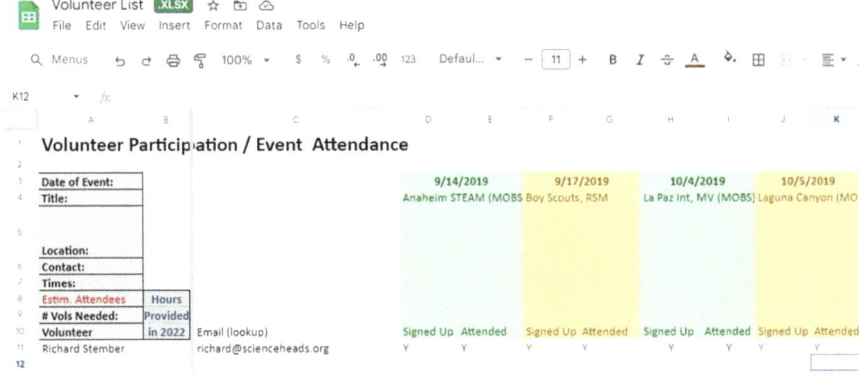

Fig. 6.5 Volunteer event scheduling spreadsheet

6.3.3   Fund Raising

Running an organization requires money. Depending on the size and scope of your organization you may need hundreds or thousands of dollars to conduct outreach events. Some of this money will be needed for essentials like general liability insurance and office supplies. Money may also be needed for printing brochures and flyers, buying advertisements, equipment, and supplies.

Start by creating a budget. A simple spreadsheet that closely estimates the money you will need broken down into categories will suffice. Not only will this help you stay within budget, but it also helps organizations that may be interested in funding your activities. Knowing how every dollar is spent is important to any business. Funding organizations are particularly concerned that the money donated is being used for the purpose intended.

Creating a budget need not be hard. Ask your bookkeeper of accountant about the "accounts" that they track. Use those accounts as the categories in your budget spreadsheet. Then estimate how much money is needed for each category. If you have historical data (previous years' profit and loss statements for example) those numbers provide a good baseline. Figure 6.6 shows an example budget for just one chapter of Science Heads Inc.

This spreadsheet can also tell you if your organization is staying within budget. Excess spending may need to be addressed during the fiscal year (FY). Funds not spent in one category may need to be moved to another to keep the organization on target. These are the types of issues that the organizations' board of directors would typically address.

There are many ways to raise funds for a non-profit organization including:

- Membership dues.
- Subscriptions.
- Donations from individuals and organizations.
- Sales of products and services.
- Grants

Non-profits can solicit donations from both individuals and companies. For specific projects and campaigns there are also crowdsourcing sites that can help:

Site	Link
DonorBox	https://donorbox.org/
Feathr	https://www.feathr.co/
Fundly	https://fundly.com/
Galabid	https://www.galabid.com/
Givelify	https://www.givelify.com/
Give Butter	https://givebutter.com/
GoFundMe	https://www.gofundme.com/
Thankview	https://www.thankview.com/

SCIENCE HEADS
SOCAL CHAPTER BUDGET DRAFT
2019/20 FY

	2019/2020 Budget	7/1/2019 - 1/31/2020 Actual	Difference
Income			
Donations			
General Fund	9,000.00	5,705.00	(3,295.00)
Project Specific Donations	5,000.00	450.00	(4,550.00)
Event Income	5,000.00	1,360.00	(3,640.00)
Total Income	19,000.00	7,515.00	(11,485.00)
Expenses			
Advertising & Marketing	700.00	286.38	413.62
After School Program	500.00		500.00
Conferences	300.00		300.00
Equipment	3,500.00	3,309.05	190.95
Fundraising Expenses	200.00	29.00	171.00
HAB Expenses	1,500.00	1,417.22	82.78
HAB Equipment	300.00	564.28	(264.28)
HAST-C Sessions	4,400.00	852.86	3,547.14
Insurance	2,000.00	325.00	1,675.00
Meeting Expenses	200.00	109.53	90.47
MOBS Equipment	250.00	28.81	221.19
MOBS Maintenance	350.00	11.00	339.00
Office Expense	400.00	83.71	316.29
Outreach Expenses	1,000.00	59.46	940.54
Parade Expenses	50.00	37.15	12.85
Storage	2,200.00	1,408.00	792.00
Taxes	25.00		25.00
Meals	0.00	167.25	(167.25)
Volunteer Appreciation	0.00	168.20	(168.20)
Total Expenses	17,875.00	8,688.70	9,186.30
Net Income	1,125.00	(1,173.70)	(2,298.70)

Fig. 6.6 Example of a simple budget

Grants are available from non-profit foundations and some government agencies. Grant applications may be accepted only during specific months of the year or when government funding becomes available. Some organizations only accept applications from non-profits that are invited to apply. Each organization will differ in its rules and procedures.

The best way to find grant opportunities is to search for foundations that you know have an interest in what you do. There are subscription-based services that maintain databases designed to make the search easier. Figure 6.7 lists a few sites that lists grants specific to astronomy outreach efforts.

SITE	LINK
Asterion Foundation	https://asterionfoundation.org/
NASA	https://nspires.nasaprs.com/
National Science Foundation	https://new.nsf.gov/funding
Society for Science	https://www.societyforscience.org/outreach-and-equity/stem-action-grants/
STEMfinity	https://stemfinity.com/pages/stem-grants
STEMGrants	https://stemgrants.com/stem-grants-for-k-12-nonprofits/

Fig. 6.7 STEM grant opportunity sources

6.3.4   Communications

Keeping a team well organized requires a lot of communication. Sending emails to volunteers, hosts, venues, supporters, and other people who have a role in your organization may often require a substantial amount of time. It can seem like this is half the job of running a non-profit organization.

There are methods that can help reduce your labor. Most are based on common sense. The following are tips for communicating with school staff and volunteers about an event:

Tip 1 – Include all your contact information in your email signature block. The email recipients can hit the reply button if they want to send you a response by email. But if they urgently need something, and your telephone or fax number is not in the email, they may have to search for it - decreasing their efficiency and possibly delaying a response that you need. Create a standard signature block using your email software. The block should include all of the ways that you can be contacted.

Tip 2 – Don't expect an immediate response to emails from volunteers and school staff. Many volunteers have real jobs and all of us have other obligations. Emails often get ignored, forgotten, or read later when it is more convenient for the recipient. Send out emails with plenty of lead time. If you need a volunteer's help don't wait until two days before the event to ask. Send the request early. If possible, allow them a couple of weeks to respond.

Tip 3 – Don't worry about pestering your volunteers with follow up emails. Weekly updates and reminders to your team will be appreciated. Figure 6.8 shows the format that Science Heads Inc. uses for its event notification emails.

Tip 4 – Teachers and school staff are always busy. Emails sent in the middle of the week may not get attention or even noticed. This is especially true at the beginning of the school year. When a new year starts, school staff are often just treading water dealing with new schedules, new students, and new procedures. If you must send communications early in the school year – do it before the school year begins. Teachers are often at the school one or two weeks before the students arrive. This is a good time to introduce yourself, your services, and start scheduling events.

Tip 5 – When you receive a request for an event respond with all the information that you know is important. Be sure to ask the sender the important questions:

- What days of the week are best? What times are best?
- What is the objective of the event? Topics they want covered? Grade levels and ages?
- Is proof of insurance required?
- What language should be put on the Certificate of Insurance?

The requester may not immediately know all the answers – but they now realize what you need from them.

Tip 6 – Astronomy related factors like sunset and lunar phase are rarely on the requester's radar. They are only considering their school's calendar and things that they are familiar with. Provide them with the information that you know is important right up front. If they suggest a date that falls on the full moon – let them know that date is bad for lunar viewing. If Jupiter doesn't rise during the time requested but will the following month, tell them – they may opt to wait.

Tip 7 – Send out event recruitment emails early and as soon as possible. As soon as a date is set for an event, add it to the list. Volunteers' plans often change. They may initially indicate that they can help only to drop out before the event begins because of last-minute obligations.

Volunteering is important but there are always higher priorities. Sending out the event notifications weeks in advance will provide you the extra help and time in case one or two volunteers drop off the list.

Hello Fellow Science Heads Volunteers:

   The new school year has started, and we are receiving plenty of event requests. Below are the outreach events currently scheduled for the next two months. Please respond to this email if you are available to help. As always, we appreciate you and your commitment to supporting STEM education.

   1. Family STEAM Night at Laguna Niguel Elementary
      Date: Saturday, September 23rd, 2023
      Time: 6:00 pm – 8:00
      Setup Time: 5:00 pm
      Location: 1240 Main Street, Laguna Niguel, CA (link)
      Event Plan: Mobile observatory and standalone telescopes.
      Volunteers Needed: 4+
   2. ...

Fig. 6.8 Example of call for volunteers email

6.3.5 *Safety*

The safety of your audience, staff, volunteers, and yourself should always be the highest priority. Several factors common to astronomy events can raise significant safety concerns including:

- Operating at night when it is difficult to see.
- Excited and enthusiastic children running around.
- A requirement to climb steps and ladders when using telescopes and entering and exiting the observatory.
- Looking at the Sun during daytime solar viewing events.
- Volunteers driving vehicles where people are present.
- Obstacles like poles, walls, and basketball hoops present on blacktops.
- Use of electrical outlets, extension cords, and generators.

Each of these concerns can pose a significant risk to volunteers, staff, and audience members. The Outreach Astronomer should have plans and procedures in place to mitigate each risk. Clear instructions should be provided to the volunteers at an event. Safety cones, tape, and line management techniques should be employed when needed. Safety plans should be reviewed with the host and venue before the event.

Science Heads Inc. regularly uses the following safety precautions at its events:

- Power cords are covered or taped to the ground to reduce tripping hazards.
- Obstacles are surrounded with safety cones.
- Generators are placed well away from interior spaces and areas used for the event.
- Setup and tear down times are strictly enforced to prevent vehicles being driven near event attendees.
- An event volunteer is tasked with directing traffic on the blacktop.
- Volunteers are instructed on safety procedures for viewing the Sun.
- Ladders and steps are illuminated with red lights.

Protecting Minors

Organizations that work with children should have child protection protocols in place. Many states now have implemented laws requiring that volunteers take *Child Abuse Prevention* training (also referred to as *Mandated Reporter* training). Outreach Astronomers should become familiar with the local state requirements and arrange to take the required training if necessary.

Science Heads Inc. has a written policy on protecting minors that aligns with local and state mandates. Among other procedures, the Science Heads Inc. policy requires:

- A board member who has received Mandated Reporter training must be present at every outreach event.
- A minor is never alone with just one volunteer or staff member. Two adults must always be present.
- Background checks must be performed for all board members and volunteers per legal state requirements.

- All safety issues regarding a minor must be immediately reported to the event host, and the Science Heads Inc. Mandated Reporter that is present at the event.

Written child protection procedures protect not only the children at your event but also your staff, volunteers, and yourself. Before the start of an event ask about the procedures the event host and venue have in place. Identify who are the *Mandated Reporters* present at the event. Immediately share any concerns with the appropriate individuals. Figure 6.9 shows a Pre-Event Requirement Checklist used by Science Heads Inc.

EVENT REQUIREMENTS CHECKLIST

Venue: _____   Date of Event: _____

Contact(s): _____   Tele: _____

Email(s): _____

Number of Volunteers Needed: _____   Notes : _____

Event Activity Plan: _____

_____

BUSINESS REQUIREMENTS:

Liability Insurance Requirement: $ _____   Background Checks (BC) Required (Y/N) __

Workers Comp Insurance Waived/ Required (circle)   Date Waived _____

SH Individuals Participating in Event   BC Expires: MR (Y/N)   Role

_____ _____   ____ _____

_____ _____   ____ _____

_____ _____   ____ _____

_____ _____   ____ _____

_____ _____   ____ _____

Notes: _____

_____

Continue on back as necessary.

Fig. 6.9 Event requirements checklist

Health Precautions

During the pandemic, organizations and individuals found it necessary to implement COVID-19 mitigation procedures. Science Heads Inc. implemented standard procedures that were recommended by the Centers for Disease Control (CDC). These procedures included wearing face masks, maintaining distance, and providing adequate air circulation to keep our guests and volunteers healthy. Volunteers were required to self-report and cancel participation if they felt ill before an event.

At every event, face masks were made available to both volunteers and guests. Volunteers were instructed on the CDC recommendations. Science Heads Inc. took the opportunity to model best practices for our guests. The procedures were not enforced for guests, but volunteers were trained how to answer related questions. Science Heads Inc. took its role as a science-based informal education organization very seriously.

COVID-19 remains a concern and is still responsible for thousands of deaths each year in the United States. Continuing to take precautions, particularly when inside, is something that the Outreach Astronomer may want to consider. The CDC recommends continuing vigilance and staying informed. Information is regularly updated and posted at the CDC COVID-19 webpage: https://www.cdc.gov/coronavirus/2019-ncov/prevent-getting- sick/prevention.html.

6.3.6   Public Relations

Every organization that relies on public support and goodwill needs a public relations (PR) plan. A PR plan is different than marketing and advertising plans – although there can be overlapping aspects. You can think of a PR plan as a long- term strategy while marketing and advertising plans are short term.

The objective of a PR plan should be well defined and achievable. It is often designed to improve and support the relationship between your organization and the community that it serves. When you first start an organization there is no relationship. Over time you will develop relationships with both individuals and organizations in the community.

To go beyond individual relations, a concerted effort is needed to send the appropriate messages to the broadest set of members of the community possible. Examples of PR efforts include newspaper articles about the good work that your organization does, radio interviews, and TV appearances. Over time the community will recognize the value of your organization. This becomes the bedrock your organization needs for continuing financial support and volunteers.

The common technique called SWOT analysis (Strengths, Weaknesses, Opportunities, and Threats) is a good way to define PR objectives. The process usually involves enlisting a team of knowledgeable people who can help you define specific aspects about your organization and your brand:

Strengths – What currently is done well? What do people like about your services? What resources do you have readily available? What are the best things about the organization?

Weaknesses – What is not done well? What things need to be improved? What communications are not done effectively?

Opportunities – How can your communications be improved? What avenues are available but not used? Who can you be better communicating with?

Threats – What are the barriers to your efforts? Limitations; counter- productive issues?

After brainstorming and producing a SWOT analysis, your team can create an action list by addressing each opportunity and threat. How can each be mitigated and addressed? What is required? What are the costs involved?

Lastly your team should prioritize the action list. What actions would be most effective and beneficial to pursue? Are there alternative ways to address the issues?

By narrowing down the list and setting priorities the objective of your PR plan becomes obvious. This process can be remembered using the acronym SMART:

Specific – Be specific with the objective - not overly broad.

Measurable – How will you measure the degree of your success?

Attainable – Be realistic in your goals.

Relevant – The identified actions must be ones that will make a difference.

Time – Identify how long it will take to achieve the objective.

There are many different tasks that can help you achieve your PR objectives. Some of the more common tasks are:

- Regularly sending out press releases about your events.
- Submitting articles to local newspapers and other media outlets.
- Sending newsletters to your mail list of supporters and followers.
- Invite media outlets to cover events.
- Involve community leaders and well-known personalities.

Writing a Press Release

Writing a press release is as simple as remembering the five W's we learned in grade school: Who, What, When, Where and Why. It's best to keep the press release as short and simple as possible. There is no need to write an entire newspaper article after all that is the job of journalists.

Figure 6.10 shows a simple format for a press release. Always remember to include your contact information so the recipient can contact you to get more information if needed.

Identifying who to send a press release to will require a bit more work. Television, radio stations, and social media are the most popular outlets for news currently. The circulation of newspapers and magazines has dropped in recent years but do still enjoy significant readership. The first step is to identify your target media outlets. Focus on media that will reach the intended audience.

Contact: Richard Stember
Telephone: (xxx) xxx-xxxx
Email: richard@scienceheads.org
Website: www.ScienceHeads.org

FOR IMMEDIATE RELEASE

RIBBON CUTTING EVENT – MOBILE OBSERATORY
Introducing an exciting new STEM resource for Orange County, CA.

Lake Forest, CA – Science Heads, Inc., a local non-profit that supports STEM (science, technology, engineering, math), education is introducing its newest resource – a mobile astronomical observatory. Named after America's first space satellite, the Explorer1 mobile observatory is being made available to local schools, libraries, museums, and public venues to offer students and the public amazing views of the solar system. Science Heads Inc. will hold a ribbon cutting event at the Mission Viejo Library on April 10, 2017. The event starts at 10:00 am and runs until 12:00 noon. The public is invited, and attendance is free. Using specialized solar telescopes, the public will be invited to view the Sun and enjoy STEM activities for children of all ages.

Science Heads Inc. is a 501(c)(3) public benefit non-profit organization based in Lake Forest CA. Its mission is to support STEM education and raise science literacy in the local community.

Fig. 6.10 Example press release

You may want to target a specific geographic area, age group, certain demographics, or the widest coverage possible. For the latter, TV and radio are good outlets to use. Many stations have community calendars that list local events on the air. They also accept public service announcements (PSAs).

For narrower objectives, consider the audience reached by the outlet. Some newspapers and websites are published specifically for certain geographic locations and communities. Many newsletters and websites cover specific topics – science, raising children, home schooling, kids' activities. All of these could be good targets to advertise an astronomy outreach event.

Create a list of media outlets that will help you achieve your goals. Visit their websites, read their published materials, and find the names and email addresses of the journalists who may have an interest in what you are doing. Also look for the names of editors of the departments that cover education events and news. Journalists also often have a presence on social media sites. Search on X (Twitter), Facebook, TikTok, Instagram, Discord, etc. by employer or occupation to find them.

Keep your lists up to date. Identify which media outlets and journalists were responsive, covered, or showed up at your event. Keep in contact with these people, add them to your lists, and continue working toward your PR objective. The process requires an on-going effort.

Chapter 7
Assessing Effectiveness

Chapter 4 identified the importance of defining long term objectives and short-term goals. Knowing what you are trying to achieve at an event, or over the long term, is important to your funders. So is knowing if you are making progress achieving the objective. Progress only becomes known when measurements are taken, data collected, and analyzed.

Measurements are also a prerequisite to making improvements and course corrections. Both should be expected and welcomed when running STEM programs. Knowing what you do well and what you need to improve will you help you be more effective.

Funders want to know and understand your organization's effectiveness. Collecting metrics after an event and over time can be useful when applying for grants and asking for donations. There is a lot of competition for grant money; documenting your organization's impact and success gives you an advantage raising funds.

Ever since the National Science Foundation invented the term STEM, there has been research conducted on how to best assess the effectiveness of STEM programs. These studies have identified several approaches and tools that can be employed for this purpose. Which tools to use depends on the duration of the program being evaluated. The tools for short term programs, such as a single outreach event, tend to be simpler in nature and easier to use. Long term programs, like after school programs, tend to require a greater effort and more complex techniques.

The different types of assessments that can be made are:

Summative – At the end of the program were the goals achieved?
Formative / Implementation – While the program is active is it following the plan?
Formative / Progress – While the program is active is progress being made toward the stated goals?

© The Author(s), under exclusive license to Springer Nature Switzerland AG 2024　　127
R. Stember, *Share the Universe*, The Patrick Moore Practical Astronomy Series,
https://doi.org/10.1007/978-3-031-53495-9_7

7.1   Assessing Events

A typical goal of a one-time event may be to raise interest in the subject matter or teach students about a specific topic. Assessing the progress achieved after an event can be done using simple surveys and questionnaires. Handing out a card or form to the students as they leave the event can provide useful information for this purpose.

Polling professionals will warn you that there can be implicit biases in how questions are asked, who is surveyed, and who tends to respond. Care should be taken when crafting language to make sure a desired response is not unintentionally suggested. Using words that convey negativity or positivity can lead to bias. Asking "Did you enjoy the event?" vs. "Would you recommend this event to others?" can produce very different responses even though the questions are very similar. The word "enjoy" has a positive connotation. Using it may unintentionally inform the people it's the response you favor.

Respondents with strong opinions are also much more likely to return a response. On the other hand, people who don't like complaining by nature may be hesitant to mail the survey back. Professionals use mathematical weighting techniques to normalize data collected to adjust for response biases.

Even requiring that the surveyed person purchase a postage stamp can add a bias to the data. Buying a postage stamp involves convenience and financial factors. Not everyone goes to the post office or even buys stamps anymore. Pre-applying postage to a survey response card can help eliminate this potential bias.

In-person interviews can also be employed at events. In these cases, care should be taken to, as much as possible, randomly select people to be interviewed. The sample should be representative of the audience. Note that if you are interviewing children, be sure to first get approval from their parent or guardian.

Don't forget to survey teachers at the school and other school staff involved. They often have good insight and open channels of communication with their students. A student that you approach may be hesitant to give an honest response. But they may have a more open communications with their teachers.

When surveying children keep the questions unbiased, short, and concise. Figure 7.1 shows a sample survey for students.

Surveying teachers and staff can give insight to help you achieve your goals. Figure 7.2 shows a sample teacher survey.

Science Heads Inc. uses a different format for assessing public events at parks and libraries. Figure 7.3 shows one of these post event surveys. An assessment

POST EVENT SURVEY FOR STUDENTS

1.   Do you like astronomy? Y/N ____
2.   What did you think of tonight's event _____ 1= Did not enjoy it ... 5=Loved it!
3.   How much did you learn tonight ? _____ 1=Nothing new ... 5=A lot!
4.   Would you attend another event like this one? Y/N _____

Fig. 7.1 Sample interest survey of students

POST EVENT SURVEY FOR TEACHERS

1. How well did the activities align with your lesson plans? ____1-5 (5 being best)
2. How effective were we in engaging with your students? ____1-5 (5 being best)
3. Rate the student's interest in the activities. ____1-5 (5 being highest)
4. Rate the overall effectiveness of the event. ____1-5 (5 being highest)
5. What suggestion do you have for improvement: _____

_____
_____
_____

Fig. 7.2 Sample teacher survey

POST EVENT SURVEY

Event Venue: _Laguna Coast Wilderness Park_   Date of Event: _8/24/19_

Name of Person Completing Survey (Optional): _____

Email (Optional): _____   Phone (optional): _____

Instructions:

Please provide a rating from 1 to 5 (5 being the best/highest, 1 being the worst/lowest) for each question below:

1. How well did Science Heads met your overall expectations and needs for the event?

   (1-We missed the mark, 5-Fully met expectations/needs) __5__

2. How engaged was the audience in the activities provided by Science Heads?

   (1-Not interested, 5-Fully into it) __5__

3. Was the activity provided by Science Heads age appropriate?

   (1-We missed the mark, 5-Fully age appropriate) __5__

4. Were the Science Heads personnel connecting with the audience appropriately?

   (1-Talking over their heads, 5-Fully engaged and understood) __5__

5. Were the logistics of the activities (e.g. power, line management, safety, etc.) handled appropriately?

   (1-Many problems, 5-No issues at all) __5__

   Please list any areas of concern: _____

   _____

Fig. 7.3 Example of non-school event survey

survey should be written with the objectives of the host or venue in mind. Their goals are typically very different than that of teachers. A public venue may be more interested in providing an activity that is appreciated by their community. Does it draw a significant audience? Was the topic of interest? They may have specific concerns about how the event was conducted. Does it align with park regulations and goals of the municipality?

7.2 Assessing Long-Term Programs

Measuring progress towards achieving long-term objectives is often more involved. Assessing long term programs can involve measuring both short term and long-term goals. The techniques used for measuring short term goals are like those used for one-time events. For example, surveys measuring if the students were engaged in the activities. Was there interest in the subject matter? Are the materials used well designed? Do they align with the teacher's lesson plans?

Measuring longer term goals may require collecting both longitudinal and demographic data about the audience. This can include data about racial and socioeconomic makeup of the audience. Comparing test scores before the program started to test scores after the program ended may be necessary. Other tools such as student journals, interviews, and third-party observations can also be useful for long-term program assessments.

Analyzing the data and completing the assessment will require writing a thoughtful and comprehensive report. The report should be organized in a way that helps the reader make sense of the data and the conclusions reached. Charts, tables, and testimonials included in the report can help illuminate and explain the results. Typical sections of an assessment report are:

- Introduction
- A description of the program and the issues it addresses.
- Objectives, goals, and rationale of the program.
- Constraints of the program.
- Description of the audience and demographics.
- Description of the methodology used.
- The data collected.
- An analysis of the data.
- Conclusions based upon the analysis.
- Summary.

Demographic data is readily available from schools, city governments, and web sites. Below is a list of web sites that Science Heads Inc. has found useful:

National Center for Education Statistics (data by school district) https://nces.ed.gov/

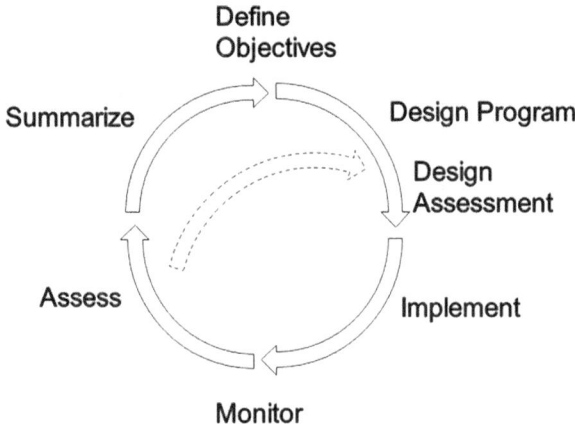

Fig. 7.4 Assessment as part of project design

Urban Institute (data by school or district) https://educationdata.urban.org/data-explorer/

U.S. Census Bureau (data and visualizations by state, county, town, city). https://www.census.gov/data.html

ZipDataMaps (data by community) https://www.zipdatamaps.com/index.php

As shown in Fig. 7.4, program assessment should be part of all projects. It should be integral to the implementation and execution of the program. Assessment is as important as are the objectives and goals of the program. Without an assessment there is no way to know if the objective and goals were met and to what degree. Without an assessment, improvements that are needed may never become known.

The National Science Foundation (NSF) provides grants for many STEM related projects both short-term and long-term in nature. To help its grantees understand how best to assess their programs, the NSF published a handbook on the topic in 2010. The *User Friendly Handbook for Project Evaluation*, was written by Joy Frechtling Westat, to help program managers design and conduct assessments for NSF funded projects. The document is available at no charge at https://www.informalscience.org/2010-user-friendly-handbook-project-evaluation. It is recommended reading and a good place to start learning about measuring the effectiveness of STEM programs.

Chapter 8
Hand-on Activities

Astronomy outreach events need not be limited to just viewing objects through telescopes or listening to a lecture. Many outreaches can be augmented with hands-on activities for children and adults which support the theme and objective of the event.

There is an almost unlimited source of ideas for astronomy related hands-on activities available on-line. Many can be easily adapted for use at outreach events. NASA provides many activities at its STEM related websites including:

NASA Learning Resources https://www.nasa.gov/learning-resources/
NASA For Educators https://www.nasa.gov/learning-resources/for-educators/
NASA @ Home https://www.nasa.gov/nasa-at-home-for-kids-and-families/
NASA STEM Content https://www.nasa.gov/stem-content/teach-stem-activities/
Jet Propulsion Laboratory https://www.jpl.nasa.gov/edu/teach/
NASA Goddard Activities https://www.nasa.gov/learning-resources/for-kids-and-students/
Pinterest NASA Collection https://www.pinterest.com/nasa/nasa-stem/

For its outreach programs, Science Heads Inc. has created over 80 activity kits related to different STEM subjects. Some of these kits have been provided to libraries and can be checked out on demand by library patrons.

The following activities are a few of the space and astronomy related activities frequently used by Science Heads Inc. at its outreach events. Figure 8.1 lists the activities included in this guide along with reference to framework and grade level.

R. Stember, *Share the Universe*, The Patrick Moore Practical Astronomy Series, https://doi.org/10.1007/978-3-031-53495-9_8

#	ACTIVITY TITLE	NSF FRAMEWORK	GRADES
1	Draw what Galileo saw.	PS4	2 – 5
2	What did you see in the telescope.	ESS1	2 – 5
3	Draw the Sun.	ESS1	2 – 5
4	All About Eclipses.	ESS1	8+
5	Lunar Phases.	ESS1	4 – 5
6	Solar System Distances.	ESS1	8+
7	Reasons for the Seasons.	ESS1	5 – 8
8	Constellation Identification.	ESS1	5+
9	Make Your Own Constellation.	ESS1	K – 1
10	Introduction to Spectroscopy.	PS4	12+

Fig. 8.1 Table of hands-on activities

8.1   Draw What Galileo Saw

NSF Framework Reference: PS4
Age Category: Grades 2–3

Goal:

Students are introduced to the scientific methodology employed by Galileo Galilei.

Activity Description:

Students observe Jupiter using a telescope. They may observe the four Galilean moons or one or two of the moons may be eclipsed by the planet. By comparing the drawings over time students may be able to discern that the moons are orbiting around the planet replicating the observations made by Galileo in 1612 CE.

Materials List:

Handout, Pencil, Telescope, a clear view of Jupiter.

DRAW WHAT GALILEO SAW

In 1612 CE, the Italian astronomer Galileo Galilei observed the planet Jupiter through an invention that we now call the "telescope." Over several days of viewing, he made a startling discovery: Jupiter had four moons orbiting around it. His discovery challenged commonly held beliefs at the time about the structure of the solar system. What did they believe back then?

What did you see tonight? After looking through the telescope, draw Jupiter and the position of its four largest moons within the circle below. Then in a few hours look again. What changed?

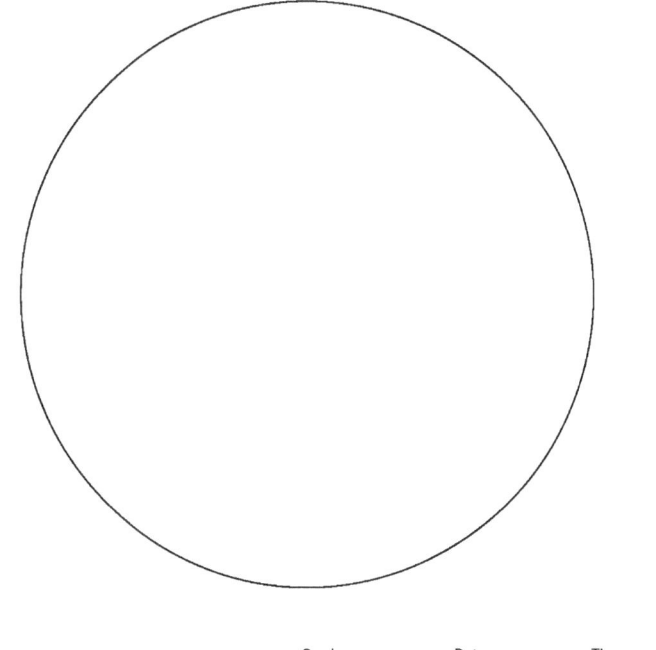

Name: _____   Grade: _____   Date: _____ Time: _____

Science Heads Inc.   Making science fun since 2014   www.ScienceHeads.org

8.2   Draw What You Saw in the Telescope

NSF Framework Reference: ESS1
Age Category: Grades 2–3

Goal:
Students learn that telescopes are needed to observe many objects in the solar system.

Activity:

Students observe objects in the night sky using a standard telescope. They may observe star fields, star clusters, nebulas, planets, or the Moon. The students learn that Galileo observed these types of objects and made several discoveries.

Materials List:

Handout, Pencil, Telescope, an unobstructed view of the night sky.

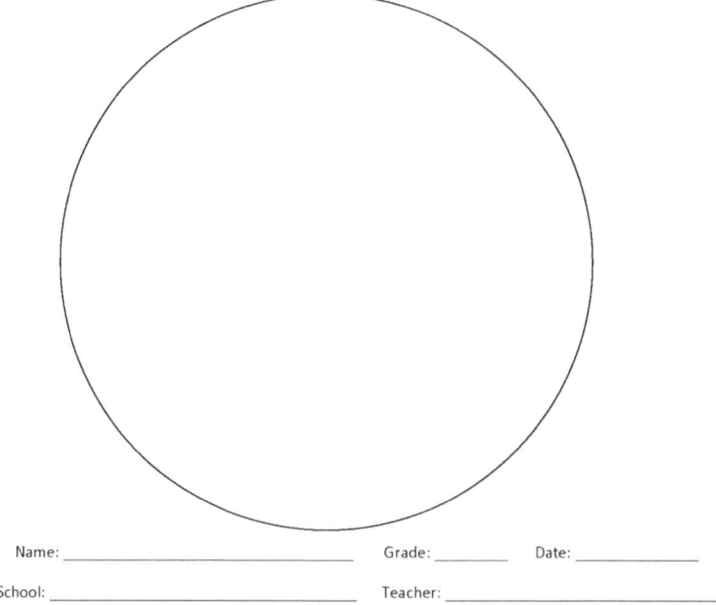

8.3   Draw the Sun

NSF Framework Reference: -
Age Category: Grades 2–3

Goal:

Students replicate the methodology Galileo Galilei employed when he studied the Sun.

Activity:

Students observe the Sun using a h-alpha telescope or a telescope fitted with a solar filter. They may observe sunspots, prominences, plages, and granulation on the chromosphere using an h-alpha telescope. Sunspots may be observed using a telescope with a solar filter.

Materials List:

Handout, Pencil, Telescope, an unobstructed view the Sun.

Draw The Sun

Over a several days in 1612 CE, Galileo Galilei drew sunspots he observed through his projection telescope. Each of these observations was made at approximately the same time of day. Galileo noticed that the spots were moving across the Sun. Up to then most astronomers thought that sunspots were moons. Galileo convinced people that they were instead something on the surface of the Sun or clouds in its atmosphere. What are sun spots? Do you know?

Draw what you saw today in the solar telescope. If you didn't see any sunspots, did you see any prominences? After you look through the solar telescope draw what you saw in the circle below.

CAUTION: Never look at the Sun through a telescope or binoculars not specially designed for solar viewing! Doing so can damage your eyesight!

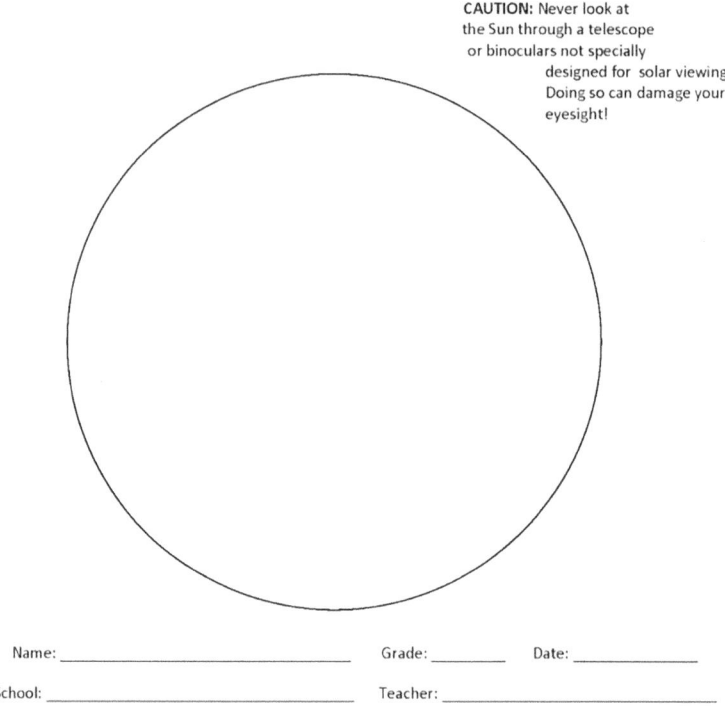

Name: _____   Grade: _____   Date: _____

School: _____   Teacher: _____

8.4   All About Eclipses

NSF Framework Reference: ESS1
Age Category: Grades 8 +

Goal:

Students learn how eclipses of the Moon and Sun occur.

Activity:

Turn on the lamp and hold the Earth stick such that it puts the Moon in the shadow of the Earth. Explain that when the Moon is in the Earth's shadow it is called a Lunar Eclipse.

Put down the Earth stick and pick up the Moon stick. Hold the Moon stick such that it's shadow appears on the Earth. Explain that this is called a Solar Eclipse.

Materials List:

Images of the Moon and Earth glued to black cardboard.

An extra bright tabletop lamp.

A small styrofoam ball painted to look like the Moon and mounted on a skewer.

A larger styrofoam ball painted to look like the Earth and mounted on a skewer.

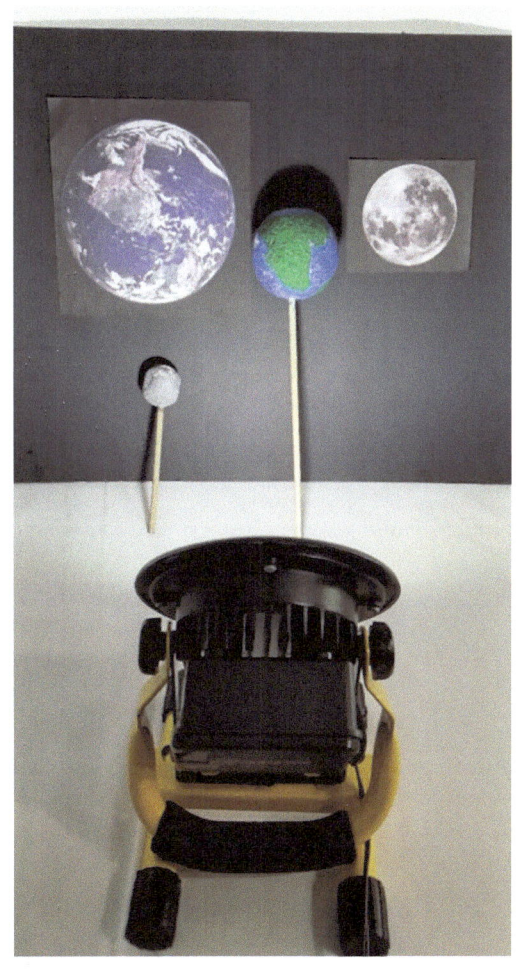

8.5   Lunar Phases

NSF Framework Reference: ESS1
Age Category: Grades 4–5

Goal:

Students learn why the Moon goes through phases.

Activity:

Before helping the participant place the board over their head, point out that the eight small balls represent the Moon during its different phases. One side of each ball is white, while the other side black. The white side represents the part of the Moon that is illuminated by the Sun which is represented by the large yellow ball.

Students then place their head inside the lunar board. Tell the student that they are now the Earth. The Moon orbits around the Earth in 29 days. By rotating their head, they can see that more or less of the Moon appears to be illuminated depending on where the Moon is in its orbit. Ask them to pay close attention to the shape of the white portion of the balls.

Explain that "waxing" means growing and "waning" means shrinking. Rotating their head counterclockwise, starting with the Moon that is nearest the Sun, the phases are: New Moon, Waxing Crescent, First Quarter, Waxing Gibbous, Full Moon, Waning Gibbous, Last Quarter, and Waning Crescent.

Materials List:

Large blac.k board with a hole cut in the center big enough to place a head through.

16 Velcro squares, 8 of which are placed evenly in a circle around the hole. One Velcro square positioned in the center line at one edge of the board.

Eight small Styrofoam balls painted one side black with a Velcro square attached.

One large Styrofoam ball painted yellow or covered in yellow felt with a Velcro square attached.

8.6   Solar System Distances

NSF Framework Reference: ESS1
Age Category: Grade 8 +

Goal:

Students will experience the relationship of distance between the planets and Pluto in our Solar System.

Activity Description:

Guests walk the distance of the twine starting at the end labeled the Sun and marked with a yellow LED lantern. They continue following the twine noting how far they have traveled to reach each planet. Beyond Mars the distances become much greater until finally they reach the end and the dwarf planet Pluto.

Materials List:

Twine cut to a length of at least 354 feet.

Velcro patches placed at the specific distances from the beginning of the twine indicated in the table below.

Signs indicating the planet names and facts with Velcro patches attached.
12 battery operated LED lanterns with dimmers.

Power cord winder or holder.

Object	Distance in feet and inches
Sun	0
Mercury	3′ 6″
Venus	6′ 6″
Earth	8′ 11″
Mars	13′ 8″
Asteroid Belt Begins	19′ 8″
Jupiter	46′ 7″
Saturn	85′ 5″
Asteroid Belt Ends	28′ 8″
Uranus	171′ 10″
Neptune	269′ 5″
Pluto	354′

8.7 The Reason for the Seasons

NSF Framework Reference: ESS1
Age Category: Grades 5–8

Goal:

Students will learn how the tilt of the Earth results in seasons and changes in temperature.

Activity Description:

Position the globe so that it is oriented such that the north pole points toward the Sun which is represented by the lamp. Explain that this is how the Earth is oriented during summer in the northern hemisphere.

Spin the globe and point out that the north pole receives sunlight all day long while the south pole is in the dark. Explain that the rest of the northern hemisphere is in sunlight more than half the day warming it more than the southern hemisphere.

Now move the globe without changing the orientation of the axis to the other side of the Sun. Explain that this is how the Earth is oriented during the winter in the northern hemisphere. Point out that the seasons are opposite in the southern hemisphere. Reposition the globe to other locations around the Sun again keeping the tilt of the Earth pointed in the same direction. Ask the student to explain the result in terms of daylight received.

Materials List:

A light bulb mounted on a wooden stand and encapsulated in a patio light globe

A small globe of the earth.

8.8   Constellation Identification

NSF Framework Reference: ESS1
Age Category: Grade 5

Goal:
Scouts and students will learn to identify several well-known constellations.

Activity Description:

The students place one black cup on the LED tabletop lamp and guess the constellation. They can compare their guess with the key provided at the activity table.

Materials List:

Tabletop LED lamp.

Multiple black cups with pin holes in the shape of well-known constellations or asterism.

8.9   Make your Own Constellation

NSF Framework Reference: ESS1

Age Category: K – 1, Girl Scout Daisy's

Goal:

Students will learn stories that ancient people told about constellations. Daisy scouts will work towards meeting their Astronomy badge requirements.

Activity Description:

Read stories that were told by ancient peoples about common constellations.

Students and scouts then can create their own constellations and stories using the materials provided.

Materials List:

Chalk of various colors.

Black construction paper cut into roughly 6″ x 9″ rectangles

Small star stickers.

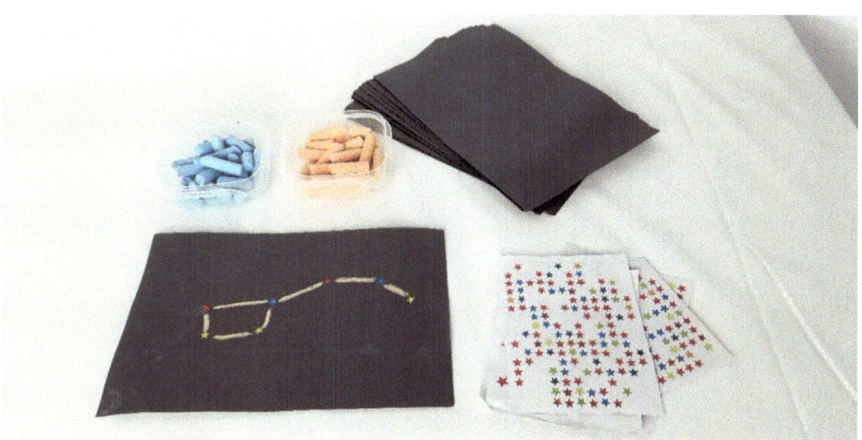

8.10   Introduction to Spectroscopy

NSF Framework Reference: PS4
Age Category: Grade 12+

Goal:

Students will learn that different chemicals, when excited, emit light at different wavelengths. When this light is passed through a prism or grating, lines become evident. The pattern of lines is specific to the chemicals present.

Identifying these patterns help scientists identify the composition of unknown materials.

Activity Description:

Explain to the student how to look through the star spectrometer. Turn on the fluorescent bulb and have the student look for the excitation lines produced by the bulb. Have them note the wavelength reading for one or two of the lines seen.

Turn off the fluorescent bulb and turn on the Niobium coated incandescent bulb. Look for the lines produced in this spectra. Note that the excitation lines have a different wavelength.

Materials List:

Star spectrometer.

Two light fixtures.

One fluorescent bulb.

One Niobium coated candescent bulb (GE Reveal).

Power strip.

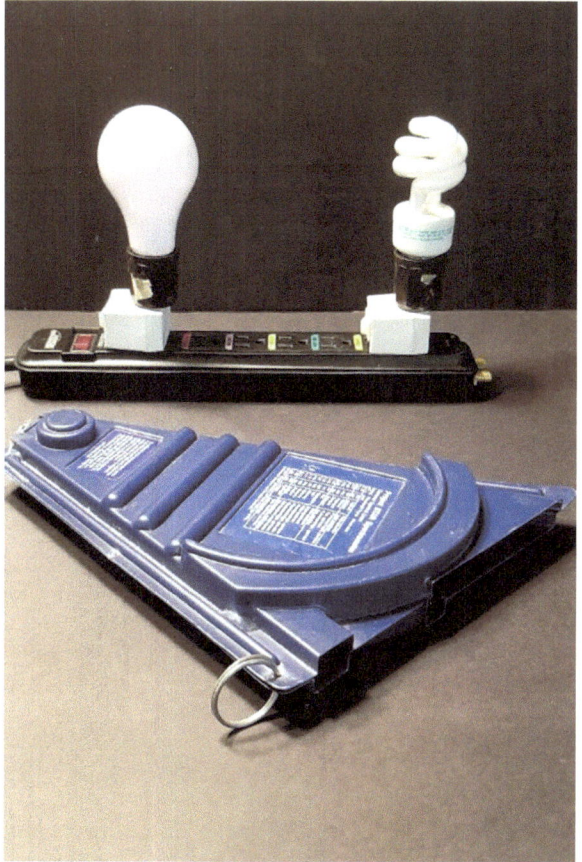

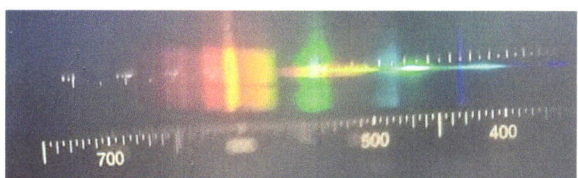

Chapter 9
Final Word

Outreach Astronomy is key to informing the public. My goal is to maximize the effectiveness of outreach efforts by "professionalizing" what we do as informal educators. This can only be accomplished by understanding how learning occurs, what children are being taught in school, and how to augment what formal and informal educational organizations are doing. Working with teachers, librarians, museum staff, and like-minded people at youth serving organizations, offers opportunities to reach a wide audience and raise science literacy.

During my years of doing astronomy outreach I have learned many truths including that being prepared with tools, materials, and best practices for working with typically underserved audiences can have a profound impact on the individuals involved. There are many resources available to the outreach astronomer for this purpose. Audio files, 3D models, braille books, and other resources are readily available to assist in the effort.

How to use these tools and techniques can be found in on-line training videos from STEM support initiatives and organizations.

Applying the techniques and ethos of Informal Interpretation can help outreach astronomers present programs that are both memorable and educational. Showing respect for other ideas and different cultures can open the door to many fruitful conversations.

It's worth noting that the science of astronomy is likely to go through a major revolution in the years to come. The *standard model* that is used to explain how the Universe formed is being challenged. Data collected by the James Webb Space Telescope (JWST) has already raised questions about its origin and formation. And the day that Dark Matter and Dark Energy are understood will be the day that modern science abruptly changes.

We know quite a bit about baryonic matter. But it only accounts for 4% of the composition of the universe. What remarkable knowledge will be gained when we finally understand the remaining 96%? It could fundamentally change everything

© The Author(s), under exclusive license to Springer Nature Switzerland AG 2024   149
R. Stember, *Share the Universe*, The Patrick Moore Practical Astronomy Series,
https://doi.org/10.1007/978-3-031-53495-9_9

that we know about science. Among the general population this could create a great deal of apprehension and skepticism.

Two favorite questions I am frequently asked during an outreach event are "Do you believe there is life elsewhere?" and "Do you think that aliens have visited our planet?"

As an optimist and a scientist, I want to believe that life does exist elsewhere. Afterall, the probability is high that conditions for supporting life will be found somewhere. NASA is actively looking for these environments on Mars, Europa and elsewhere.

The second question may appear silly on its face. If an advanced alien civilization had developed the technology needed to travel the great distances of the universe, why would they want to visit Earth? And why not make their presence known? Besides cataloging species on our planet, what would they learn, and what would they gain by visiting us?

But both questions are in fact evidence of a mind at work. The person asking these questions is considering the possibilities. They are using their imagination which is one of the foundational tools of modern science. Nobody has directly seen an atom, but we can imagine that it exists. Einstein wrote that he imagined traveling on a photon when he developed his theory of Special Relativity.

The discussion that follows with the person asking these questions can open the door to learning. It's a discussion worth having for that very reason.

The role of the Outreach Astronomer is to facilitate learning and critical thinking. No question is unworthy of hearing and discussing.

The NSF recognized in 2001 how important Science, Technology, Engineering, and Math are to our nation and its citizens. The scientists at the NSF believe that the public's understanding of STEM is key for our society's success. Learning about STEM subjects helps people:

- Think critically.
- Solve problems.
- Evaluate information.
- Be creative in thought and action.
- Achieve independence in their lives.

The role of the Outreach Astronomer is to facilitate informal STEM education and fortify our fellow citizens with the intellectual tools that they need. It's not just a matter of knowing what stars are made of, or how they form from nebulas; it is demonstrating how rational thinking can lead to solving the problems we face as individuals and societies. It's about understanding and trusting the power of our minds and supporting solutions that come from that effort.

Sadly, the American public currently has a very limited and simplistic understanding of how science works. The idea that a researcher is removed from what is important to society is antiquated and naïve. The mRNA vaccines that became available in 2021 to fight Covid-19 came about because of nearly 30 years of research. Hundreds of researchers labored for decades, building upon successes and failures.

One recent study estimated that 20 million lives were saved in the just the first year the mRNA vaccines became available.

But this is not the story that the public came to understand. In the decades that it took to get to where we are today, there were countless scientists involved, many successes and failures, and extraordinary discoveries and breakthroughs. All of this effort, though, was boiled down to the misconception that the mRNA vaccine appeared overnight and then was immediately approved by the FDA. Is it no wonder this misperception led to public skepticism and rejection about the vaccine?

The truth is that science provides many solutions to problems vexing the world: pollution, hunger, disease, climate change, and more. Many of these could be solved today if people had trust in science and the will and economic power to apply what we know. Trust and understanding can lead to marshalling the political and economic power needed to address these problems. So many lives could be improved, so many lives saved.

Outreach astronomers are the tip of the spear of this educational effort. It's not just astronomy. You represent all sciences. You show the value of rational thought, critical thinking, and intellectual honesty. Science is the best tool humankind has available to advance knowledge and better the human condition. I invite you to go forth, educate, respect others, and learn from and enjoy the effort.

Clear Skies!

Bibliography

Adams, Henry. (1905). The Education of Henry Adams. Digireads.com Publishing.

UNSESCO Institute for Statistics, http://uis.unesco.org/en/glossary-term/formal-education. Accessed June 28, 2023

Falk, J. and Dierking, L. (2010). The 95 Percent Solution. American Scientist Sigma Xi, Vol 98.

The Numbers. *www.the-numbers.com. Accessed June 28, 2023.*

Jeffs, T. and Smith, M. K. (1997, 2005, 2011). What is informal education?, *The encyclopedia of pedagogy and informal education*, https://infed.org/mobi/what-is-informal-education. Accessed June 28, 2023.

H. Gardner. (1999) Intelligence Reframed: Multiple Intelligences for the 21st Century. Basic Books, NY, NY.

Prodigy. The Ultimate Guide to Teaching Methods for Modern-Day Teachers, https://www.prodigygame.com/main-en/blog/teaching-methods/. Accessed June 28, 2023.

I. Bouchrika, Research.com. Teaching Methods and Strategies: A Complete Guide to Techniques, Styles & Trends, https://research.com/education/teaching-methods-and-strategies-guide. Accessed July 3, 2023

Center for Advancement of Informal Science Education. https://www.astc.org/resources-and-learning/caise/. Accessed June 28, 2023

National Informal STEM Education Network. https://www.nisenet.org/. Accessed June 28, 2023.

National Association for Interpretation. https://www.interpnet.com/. Accessed on June 28, 2023.

Tilden, F. (2007). Interpreting our Heritage, Fourth Edition. University of North Carolina Press. Chapel Hill, NC.

S. Ham, (1992) Environmental Interpretation: A Practical Guide for People with Big Ideas and Small Budgets, North American Press, Golden, CO.

National Park Service. www.nps.gov/idp/interp/102/infinterp.htm. Accessed June 28, 2023.

Schramm, Wilbur (1971). The Process and Effects of Mass Communication, Revised Edition. University of Illinois Press, Urbana, IL.

National Science Foundation. Science and Engineering Indicators, 2002, https://eric.ed.gov/?id=ED463970, Accessed June 28, 2023.

Kuhn, T. (1996). The Structure of Scientific Revolutions, The University of Chicago Press. Third Edition.

Willard, Ted (eds) (2015). The NSTA Quick-Reference Guide to the NGSS K-12. NSTA Press, Arlington, VA.

National Research Council (2012). A Framework for K-12 Science Education: Practices, Crosscutting Concepts, and Core Ideas. Committee on a Conceptual Framework for New K-12

R. Stember, *Share the Universe*, The Patrick Moore Practical Astronomy Series,
https://doi.org/10.1007/978-3-031-53495-9

Science Education Standards. Board on Science Education, Division of Behavioral and Social Sciences and Education. Washington DC. The National Academies Press.

National Science Education Standards (1996), National Academy Press, Washington DC.

Girl Scout Research Institute (2022). The Impact of Girl Scout STEM Programming. Girl Scouts of the USA. NY, NY.

Krupp, Dr. E.C. (1983). Echoes of Ancient Skies – The astronomy of lost civilizations. Koneky & Koneky. Old Saybrook, CT.

Ganeri, Anita (2019). *Star Stories*. Running Kids Press, NY, NY.

USA Facts.org, https://usafacts.org/articles/30-years-after-americans-disabilities-act-one-eight-americ/, accessed on August 23, 2023.

Grice, Noreen (2012). Everyone's Universe, Second Edition. You Can Do Astronomy LLC. New Britain, CT.

Littmann, M., Espenak, F. Willcox, K. (2009). Totality Eclipses of the Sun, Third Edition. Oxford University Press. Oxford, NY.

Hearing Loss Association of America, https://www.hearingloss.org/wp-content/uploads/HLAA_HearingLoss_Facts_Statistics.pdf. Accessed August 23, 2023.

Dolgin, Elie. The Tangled history of mRNA vaccines, Scrbd, 14 sept 2021, https://www.scribd.com/document/609553964/The-tangled-history-of-mRNA-vaccines?utm_medium=cpc&utm_source=bing&utm_campaign=3Q_Bing_Search_Beta_NB_US&utm_adgroup=Dynamic&utm_term=scribd&utm_matchtype=b&utm_device=c&utm_network=o&msclkid=355cc12a7ff313798790a5d5ff65807d. Accessed September 11, 2023.

Shukla Deep. Medical News Today, Covid-19 vaccines saved 20 million lives in 1 year, https://www.medicalnewstoday.com/articles/covid-19-vaccines-saved-20-million-lives-in-1-year. Accessed Sept 11, 2023.